CAMBRIDGE COUNTY GEOGRAPHIES

General Editor: F. H. H. GUILLEMARD, M.A., M.D.

T0175985

CHESHIRE

Cambridge County Geographies

CHESHIRE

by

T. A. COWARD

With Maps, Diagrams and Illustrations

Cambridge:
at the University Press
1910

CAMBRIDGE UNIVERSITY PRESS
Cambridge, New York, Melbourne, Madrid, Cape Town,
Singapore, São Paulo, Delhi, Mexico City

Cambridge University Press
The Edinburgh Building, Cambridge CB2 8RU, UK

Published in the United States of America by Cambridge University Press, New York

www.cambridge.org
Information on this title: www.cambridge.org/9781107639263

First published 1910
First paperback edition 2013

A catalogue record for this publication is available from the British Library

ISBN 978-1-107-63926-3 Paperback

CONTENTS

ILLUSTRATIONS

MAPS

The Illustrations on pp. 11, 19, 22, 40, 47, 60, 85, 88, 95, 128, 139, 144, 156 and 190 are from photographs by Mr T. Baddeley, Manchester; those on pp. 35, 105, 110, 121, 126 and 129 from photographs by Mr A. Newstead, Chester; those on pp. 148, 149, 151 and 202 by Mr C. Coote, Sale; those on pp. 42

and 101 by Mr A. J. Martyn, Hoylake; that on p. 125 is from a photograph by Mr R. Newstead, Chester; that on p. 96 is from a photograph supplied by Mr B. R. Lucas, Winnington; and those on pp. 7, 9, 12, 14, 16, 21, 24, 34, 51, 58, 65, 69, 89, 91, 99, 107, 113, 124, 131, 133, 135, 136, 137, 138, 141, 147, 153, 161, 167, 185, 186, 187, 188, 191, 194, 195, 198, 199 are reproduced from photographs supplied by Messrs F. Frith & Co., Ltd., of Reigate.

The Illustration on p. 93 is reproduced by permission of Messrs Cammell Laird & Co., Ltd., Birkenhead; the diagram on p. 165 is based on a map kindly lent by The Manchester Ship Canal Company; and the portrait on p. 179 is reproduced by permission of Miss Gaskell.

1. County and Shire. The Word *Cheshire*. Its Origin and Mean= ing.

It has been well said that our national history is made up of local history, and that our knowledge of the history of England as a whole will be all the better if we learn something of the way in which the English kingdoms were formed. This will help us to understand the relation which our modern divisions bear to the ancient ones. These modern divisions are named counties and shires, and we call one Kent and another Cheshire. In the latter instance, we note the suffix *shire*, while in the former there is not this special ending. If we know the reason for this difference we shall then be in a better position to understand the origin of the county of Cheshire in the early days of our history.

It is instructive to take a map of England and note the divisions that end in *shire*, which it may at once be said are portions *shorn* or cut off, *shares* (for the word has the same meaning) of a larger division. Thus Cheshire, Staffordshire, and part of Lancashire were once part of Mercia, one of the great kingdoms of early English days.

Again Berkshire and Gloucestershire were formerly parts of Wessex, another English kingdom. There are other counties, however, that do not end in *shire*. Of these it may generally be said that they are the survivals of the old English kingdoms which have retained their boundaries and names from the earliest times; in some cases, as in Sussex and Kent, since the Saxons and Jutes came to settle in England.

The history of England shows that our English fore-fathers divided our land into several kingdoms, of which Mercia was one, and as Cheshire was a portion of Mercia we must first consider the kingdom and then the shire. History and geography have a very close connexion at times, and here the one subject helps to illustrate the other.

When the Jutes, under Hengest and Horsa, landed in 449 A.D., and acquired a part of Kent, the Picts and Scots, fierce raiders from Scotland and Ireland, were harrying northern and central Britain, and it was not for more than a hundred years that the Saxons or English subdued rather than colonised Middle England. When they had driven back the marauders and the Kelts into the mountains of northern England and Wales they founded the kingdom of Mercia on the borderland or marches between the advancing English and the retreat-ing Britons. In 607 A.D., Aethelfrith seized Chester, and after many struggles and changes Mercia became the chief kingdom of England, with borders extending from Yorkshire—then part of Northumbria—to Wessex and Essex, and from Wales to the Wash.

The word *shire* is derived from an Anglo-Saxon word *scir*, and meant the share or territory of a tribe. Cheshire is simply Chester shire, the territory occupied by the Chester tribe of the Mercians. The Roman name for Chester was Deva, a city standing on the river Dee, but the name Chester is of Roman origin, for *castra* is the Latin word for a military camp, and Deva was the *Castra Legionis*, the camp of the victorious XX Legion of the Roman army. In Romano-British days the name became corrupted to *Caerlleon*, for *caer*, the Keltic word for a fort, was similar to the Latin *castra*, and *lleon* was the Keltic idea of expressing legion. The Roman word for camp, however, has survived in one form or another and exists to-day in Chester. In *Domesday Book* Cheshire is *Cestre-scire*, and the county survey includes that portion of Lancashire which lies south of the river Ribble, and part of North Wales so far as the River Conway.

2. General Characteristics. Position and Natural Conditions.

From the pages of history we learn that Cheshire was an ancient and important division, the land under the military sway of a camp, which in the possession of Britons, Romans, Saxons, and Normans, played an important part as a northern and western outpost of the conquering invaders during the centuries of struggle which built up our England of to-day. Cheshire has been the scene of many noteworthy events, about which we shall read in later chapters. As a safe military base it figures in Roman history, and we repeatedly find that the Cheshire men turned the tide of battle or helped to mould the affairs of state.

Cheshire is a maritime county in the north-west of England with a short coast-line bordering that portion of the Irish Sea known as Liverpool Bay. The Mersey, a river of the greatest commercial importance, washes its northern shores, and the shipping of Birkenhead, which is really part of the great port of Liverpool, travels all over the world. As Kent is the first British land seen by most travellers from the Continent, so the coast of Cheshire, with New Brighton at its extreme tip, is the first English land seen by those who cross the Atlantic from America by the route which passes Ireland and Wales.

Lancashire, the busiest manufacturing county in England, lies to the north of Cheshire; Yorkshire, the great woollen county, touches it on the east; Derbyshire and Staffordshire, with extensive potteries and great towns engaged in the iron trade, bound it, with Shropshire, to the south; and the busiest parts of Flint and Denbigh lie to the west. Although the actual length of Cheshire shore that faces the sea is barely eight miles, the county has a coast-line of over forty miles if we include the tidal estuaries.

Cheshire, in spite of its close connexion with the great manufacturing counties, and the fact that in the north and east the cotton districts extend across its borders, is on the whole an agricultural county, being specially famous for dairy-farming and the manufacture of cheese.

There are nearly fifty per cent. more dairy cattle for every thousand acres in Cheshire than in any other county, and the milk from these cows is sufficient for all the Cheshire towns, and most of the teeming population of southern Lancashire and Staffordshire, while there is some to spare for London and the Midlands. The agricultural produce of Cheshire is carried by road and rail to all parts, for as we shall explain in a later chapter it has exceptional means of communication, owing to the numerous main railway routes passing through it.

In the centre of Cheshire, right in the heart of the agricultural district, there has existed for centuries an important trade in salt, and round its famous brine-springs and rock-salt mines has sprung up a great commercial centre, for Northwich and Winsford are busily engaged in the salt and chemical trades. Indeed we shall see that the cultivation of land and the rearing of cattle, though spread over so large an area in the county, employ only a small proportion of its large population.

The greater part of Cheshire is a fertile and un-dulating plain, well watered, and plentifully besprinkled with woodlands, but few counties possess more varied scenery. In the west are the shores, marshes, and sand-banks of two great estuaries ; in the east the rugged and picturesque moorlands, where bold escarpments and grey rocks vary the monotony of purple heather and bracken-covered slopes. We can stand in the plain and look east and south to a background of noble hills, their summits standing nearly 2000 feet above the sea, or we can mount these hills or the lower sandstone bluffs that dot the plain, and look down on fields of kine, waving corn, and green roots, and far beyond see the clouds of smoke that hang above the busy towns that look to Cheshire for a food supply from her fruitful soil.

3. Size. Shape. Boundaries.

In shape Cheshire is not unlike a teapot, the spout being the peninsula of Wirral and the handle the long narrow strip in the east which is bounded by Yorkshire and Derbyshire. The bottom of the teapot rests, not unsuitably, on the Staffordshire potteries, and the lid is represented by the northward curve of the Mersey between Northenden and Warrington.

Measured in a straight line from Hilbre Point, its most western extremity, to Woodhead in the extreme east it is 57 miles, but so great (to carry out our simile) is the height of the spout and handle, or in reality so great the southward curve of the boundary line between Wirral and Longdendale, that this straight line only passes through some 20 miles of Cheshire land. The greatest length of a line passing entirely through the county from S.W., near Malpas, to N.E., on the Yorkshire border, is 54 miles. From north to south—from near Warrington to the end of the southern promontory near Adderley—is about 32 miles. The circumference is roughly 170 miles, but the actual length of the boundary line is nearer 220 miles. The area of the Administrative County is 654,825 acres, or 1023 square miles.

Dane Valley, near Congleton

In point of size, Cheshire is the fifteenth English county, and embraces an area of about one-fiftieth of the whole of England. It is a little larger than Derbyshire, rather smaller than Staffordshire, and is below the average size of an English county.

The boundaries of Cheshire on the north are Liverpool Bay and the estuary of the Mersey, the river Mersey and its tributary the Tame. In the north-east the boundary line leaves the Tame three miles beyond Stalybridge and rises to the watershed of the Pennine Range, which it follows to the source of the Etherow at Woodhead. The eastern boundary is formed by two tributaries of the Mersey which join at Compstall, where the county is only about two miles across. These streams, the Goyt and Etherow, separate our county from Derbyshire, while the Dane, rising close to the source of the former on the eastern border, forms a considerable portion of the southern boundary between it and Staffordshire. The summits of a range of hills terminating at Mow Cop constitute another natural line, but from Harecastle to the Dee the boundary dividing the county from part of Staffordshire, Shropshire, and an outlying portion of Flint is again artificial. The Dee separates Cheshire from Denbighshire, and the estuary of this river lies between Cheshire and Flint, but from Aldford to Burton the boundary is again artificial. A portion of Cheshire land here lies west of the Dee to the south of Chester, while to the west of the city Flintshire crosses the river for some miles.

At first sight this latter divergence from a naturally-provided boundary appears strange, for there is Cheshire

land in Wales and Welsh land in Cheshire, but history
gives the explanation.   Most of the land which lies
between Pulford and Burton, and westward to Kinnerton
and Hawarden, was at one time a great tidal marsh,
forming a natural westward defence for the city of
Chester.   At Pulford and Dodleston there were castles
or forts to keep the Welsh from entering the marshes ;
and the mound and moat of the defences at the latter

Warburton Bridge, Manchester Ship Canal

village, where many a fierce fight took place between the
wild hillmen of Wales and the defenders of Cheshire, are
still visible.   The Welsh, though they frequently penetrated the first lines of defence, never succeeded in entirely
wresting this marshy land from Cheshire.

The origin of Welsh land east of the Dee is quite
different.   From Chester to Connah's Quay the Dee
runs between artificial banks, but before this direct "New
Cut" was made, it flowed further to the east, under an

old river bank which is clearly visible for many miles below Chester. The construction of the "Cut" and the embanking of the old estuary reclaimed a large tract of land in Flintshire, known as Sealand, and the county boundary still follows the line of the old river.

It has been suggested that the construction of the Manchester Ship Canal has altered part of the northern boundary; the Canal is cut straight and does not follow the old windings of the river, and as for most purposes the centre of the Canal is practically the county boundary, it is urged that portions of each county, enclosed by loops of the old river bed, should be transferred to the other. This has actually been done in one place; at Warrington the Ship Canal is now the county boundary.

4. Surface and General Features.

The succeeding chapters in the geography of Cheshire will be better understood when we have got a good idea of the chief features of its surface. The surface of Cheshire is varied, and in some parts there are hills which may almost be called mountains. Most of Cheshire is an alluvial undulating plain, but in the extreme east the ground rises rapidly to the Pennine Range and Peak of Derbyshire, which has been very well called the backbone of England.

On the northern side of the Longdendale Valley, the most picturesque part of Cheshire, the hills are nearly 2000 feet in height, and are at their tops heathery moor-

lands fringed by rocky escarpments or edges. Below, in
the valley, is the long chain of reservoirs from which
Manchester and many towns in Cheshire obtain part of
their water supply. The highest point in Cheshire is
Soldier's Lump at Black Hill on the Yorkshire border,
where these moorlands are 1909 feet above the sea.

Soldier's Lump

Extensive grouse moors cover many thousand acres of
Cheshire land southward along the Derbyshire border;
the rocky top of Shining Tor, 1833 feet high, rears above
the moors of Macclesfield Forest on the one hand and
those of Buxton on the other; the conical grass-clothed
mound of Shuttlings Low is 1659 feet high; the whale-

backed moor of Bosley Minn, 1260 feet at its highest point, and the impressive bluff of Bosley Cloud rises to over 1000 feet. Mow Cop, on the Staffordshire border, also over 1000 feet, is the furthest south of the hills in this outlying ridge of the great Peak of Derbyshire.

Summit of Mow Cop

It must not be supposed that the rest of Cheshire is level, although it is called a plain. A long ridge of land, which attains an altitude of over 500 feet at Harrol Edge and Eddisbury, starts in the north of the county with the Overton and Helsby hills which overlook the marshes of the Mersey Valley. Helsby Hill is steep and rocky on its northern slope, and stands as well as any hill in Cheshire. The high ridge runs southward through the Delamere country, most of Delamere Forest being considerably higher than the surrounding land. Eddisbury Hill is capped with remains of an ancient camp, and a little to the south the tree-crowned summit of High Billinge, though really not so high as Eddisbury, is the most conspicuous portion of the ridge from many parts of the plain. The Gowy Valley makes a slight break in the chain, but to the south again the Peckforton Hills, in several places over 700 feet high, extend as far as Bickerton, from the heathy summit of which we may look beyond southern Cheshire into Shropshire, and westward see the mountains of North Wales.

A similar but much lower ridge of high ground forms the backbone of the peninsula of Wirral ; at its northern extremity are Bidston Hill and Caldy Hill overlooking Liverpool Bay. Alderley Edge, a sandstone outcrop on the eastern side of the county, near Macclesfield, rises abruptly from the plain to a height of 604 feet, and owing to its isolation it is very prominent in the Cheshire landscape.

In ancient times the now fertile surface of Cheshire consisted of forest and marsh, and three great royal forests were reserved for sporting purposes. The Forest of Wirral has vanished, and Macclesfield Forest is now only a name for a district, but Delamere Forest is still Crown land and retains its sylvan beauty, although much of it is now under cultivation. We usually understand the word *forest* to mean a large woodland, and we talk of forest-trees, but originally it denoted an open space devoted to the chase of wild animals, and we retain this use of the word in the "deer-forests" of Scotland. All three of the Cheshire forests were probably open heaths or moors, and Macclesfield Forest to-day is mostly grouse-moor and upland pasture. Delamere Forest, however, was planted

Old Corn Mill, Dunham Park

over a hundred years ago, with the idea of providing timber for the navy. Iron ships replaced the "wooden walls of England" before the trees had grown tall and big enough to be worth cutting down, and the woodlands of the Forest remain to this day.

Two important features of Cheshire are its parks and its lakes, or, as they are locally called, meres. Cheshire was described by Speed as the "seedplot of Gentilitie," for the county has always been respected for its "most ancient and worthy families." These families naturally had their habitations, and though some of the time-honoured Cheshire names have vanished, their homes remain, generally surrounded by extensive demesnes. There are some maps of the county in which the parks are coloured green and the meres blue, and if we look at such a map we shall see that it is dotted all over with little green and blue patches. In these parks there is some fine old timber, and the trees are in many cases older than the houses themselves.

The meres and pools—some of them are natural and some artificial—frequently lie in these parks, but not in every case. When a park was enclosed it was natural that the owner should wish to have a fine sheet of water near his house, and this he secured by placing a dam at the lower end of the valley of some stream which flowed through his estate. On the Tatton estate there are two good examples of natural and artificial meres; Rostherne is a natural lake and lies in a great hollow; it is the largest in Cheshire and covers some 156 acres. Tatton Mere on the other hand, about 60 acres in extent, is artificial, but from its upper end to the dam it measures nearly a mile: it is long, narrow, and shallow, but Rostherne is over 100 feet deep in many places. Combermere in the south of the county is the next in size to Rostherne, and is 132 acres in extent.

It is said that the natural meres were originally formed in hollows caused by subsidence over the salt deposits, and whether this be so or not it is true enough that since brine-pumping became general in the salt-districts immense subsidences have occurred, and three large waters as well as many small ones have been formed. The biggest of these is still called Witton Brook, although steamers from the Weaver come into it to discharge their cargoes, but the usual name for these accumulations of

Rostherne Mere

water over a subsidence is a " flash." This is a very ancient word in Cheshire, and was used by men who wrote about the county in the seventeenth century; it comes from a French word for a depression or puddle. There are some very large flashes which are not marked at all on the ordnance map of 1842, and all of them have grown very much during the last fifty years.

At one time large areas in Cheshire were marshes, and marshy land, which was formerly tidal saltings, exists to-day near Saltney, Thornton-le-Moors, and Frodsham. But beside these tidal marshes there were great tracts of country along the river valleys which were covered with peat and were called " mosses." Nearly all of these have been drained and cultivated, and now only one or two small portions, such as a fragment of Lindow Moss, remain to show us what they were like. We shall have occasion to refer to these mosses in a later chapter ; the last of any size, Carrington Moss, which in the middle of the last century was 750 acres of heather, was taken over as a sewage farm by the Manchester Corporation and cultivated in 1886. Danes Moss, near Macclesfield, covered 600 acres fifty years ago, but now only a few peat stacks and a little boggy ground mark the site. Whitley Reed, described as one of the deepest and wildest mosses in Cheshire, was reclaimed in 1852, and now not a trace of it remains except in the names of the farms and woods in the neighbourhood.

In one place only, marshes are growing year by year. Owing to the artificial alteration of the Dee channel and the reclamation of Sealand a large tract of salting has formed between Burton and Denhall and is gradually extending its area to the north and west.

The physical aspect of northern Wirral has been altered within the last few years, for the teeming population of Liverpool has spread into Cheshire, and the towns have encroached upon the Birket marsh and the Wallasey and Leasowe sandhills. Seacombe, New Brighton, and the outskirts of Birkenhead are reaching westward, while West Kirby and Hoylake come steadily eastward to meet them.   The construction of the Leasowe embankment has completely altered the character of the once waste land behind it, and sand-dune and marsh will soon have vanished from Wirral.

5.   Watersheds.   Rivers.

There are no important watersheds actually in Cheshire, but in more than one place on the eastern border the county boundary runs along the great watershed of the North of England, the Pennine Range and its spurs in the Derbyshire Peak.   The streams on the west of the Pennines which flow down the Mersey basin enter the Irish Sea, while those to the east, the tributaries of the Trent and the Yorkshire Ouse, pass through the Humber into the North Sea.   This watershed forms for twelve miles the northern boundary of the county between Mossley and Woodhead, where the Great Central Railway tunnels through the hills into Yorkshire.

The river Mersey flows along the northern boundary of Cheshire.   It is formed by the junction of several clear mountain streams.   The Tame, one of these streams, rises in Yorkshire and soon after runs along the boundary, receiving the waters of Swineshaw Brook from the upland mosses of Longdendale ; it becomes the Mersey at Stockport.   The Etherow rises from Featherbed Moss above Woodhead, and passes through the long chain of reservoirs of the Manchester Corporation[1]. Heyden, Great Crowden, and Arnfield brooks, flowing from the millstone-grit hills

Great Crowden Brook

[1] The highest of these reservoirs is 782 feet above the sea, the lowest 486 feet. Torside Reservoir, the largest of the chain, covers an area of 160 acres, and is 84 feet deep at the lower end. In the whole series of reservoirs nearly 4,000,000,000 gallons of water can be stored.

on the Cheshire side of Longdendale, supply it with some of the purest and best drinking-water in England.

The Goyt, another clear and beautiful mountain stream, rises at Goyts Moss, on the Derbyshire border, and falls rapidly down a well-wooded and charming valley, past Disley and Marple, until it meets the Etherow at Compstall. Goyts Moss is the watershed of the Mersey and Trent basins. The combined waters of the Goyt and Etherow flow across a narrow neck of Cheshire, four miles wide, and join the Tame at Stockport. Before the combined streams have reached Stockport the water is unfortunately fouled by the effluents from mills, and after Stockport, where the Mersey passes between artificial banks through low-lying water-meadows, it is further polluted. It is not, however, until the Lancashire stream, the Irwell, or more correctly the Manchester Ship Canal, joins it at Partington that it becomes a navigable water-way. The Ship Canal will be noticed in a later chapter, and the Mersey estuary in the section dealing with the coast of Cheshire. From its source at Woodhead to Runcorn, where it really becomes an estuary, the Mersey, without taking into consideration all its windings, is about 48 miles long, and adding to this about 18 miles of estuary, we have a river flowing through or alongside Cheshire for nearly 70 miles.

The three chief Cheshire feeders of the Mersey are the Bollin, Weaver, and Gowy. The Bollin rises from the moor a mile or two east of Macclesfield, and flows through beautiful country to that town, and thence across the plain through the old-world town of Prestbury to Wilmslow, near which it is joined by the Dean, which rises in the Tors of Macclesfield Forest. It is worth noticing that the stream which flows through the stone-built town of Bollington is really the Dean and not the Bollin. The Birkin, first formed by Marthall and Pedley brooks, and then receiving Mobberley brook and the streams draining Rostherne and Tatton Meres, enters

On the River Bollin

the Bollin below Bowdon, and a mile or two further on the Bollin passes through another Bollington, this time only a village, finally entering the Mersey near Heatley.

The Weaver is the most important of the feeders of the Mersey, and throughout its length of over 50 miles is entirely a Cheshire river. It rises in the Peckforton Hills in south-west Cheshire, and flows southward to Audlem

on the southern border of the county; it then bends
sharply northward, receiving Checkley brook, Ash brook,
and Wettenhall brook, to Wharton and Winsford, passing
through two large "flashes" before it enters the latter
town. At Northwich it is joined by the Dane, and for
the last twenty miles of its course is an embanked navigable

The Weaver at Church Minshull

river, known as the Weaver Navigation, and is the
waterway for all the small steamers and barges engaged in
the great salt and chemical trades of central Cheshire.
The Weaver has no lack of tributaries. The Wheelock,
rising in Mow Cop, and passing through the now busy
little town of its own name, flows into the Dane at
Middlewich. The Dane rises in Axe Edge and Goyts

Moss and flows under Three Counties Bridge, where
Cheshire, Derbyshire, and Staffordshire join hands.
Thence it wanders through a wooded valley, still forming
the county boundary, to Bosley; here it enters Cheshire,
and after flowing through the forsaken village of Havannah,
through Congleton, Somerford, and Holmes Chapel, becomes a fair-sized stream below Middlewich, and joins
the Weaver at Northwich.

The sluggish Gowy is a stream of little importance.
It rises in low-lying land east of the Peckforton Hills, and
passes close beneath the rock on which stand the ruins of
Beeston Castle, whence it follows a straight north-west
course through low and marshy meadows. It passes no
village of importance, and beyond Thornton-le-Moors is
syphoned under the Ship Canal, and debouches on the
Mersey sands at Stanlow Point, a historical spot which
we shall read more about later.

Even if we take into account the windings of the river
Dee, its 20 miles of tidal estuary and "cut," only some
40 miles of it can be called Cheshire river. Yet the Dee
is always the Cheshire Dee and not the Welsh Dee, because that portion of its course which is famed in history
is intimately connected with Chester. The Dee, an excellent salmon river, rises above Bala in Wales, and flows
through some of the most romantic scenery of the Princi-
pality; near Shocklach it touches Cheshire, and winding
and twisting flows northward along the border, beneath
the bridge that connects Welsh Holt with Cheshire
Farndon, and where at spring tides we may see the river
flowing backwards. At Aldford it enters Cheshire and

passes the noble mansion of Eaton Hall. From here it is a fine broad river, crowded in summer with the boats of pleasure-seekers and the small steamers which ply between Eaton and Chester. At Chester it is bridged first by the Suspension Bridge, then by the ancient Dee Bridge, and

The Dee at Eccleston Ferry

then by the magnificent single-span Grosvenor Bridge across which runs the main road into Wales. Below Chester the Dee is embanked and navigable, and we shall refer to it again in the chapter on the Cheshire coast.

6. Geology and Soil.

By Geology we mean the study of the rocks, and we must at the outset explain that the term *rock* is used by the geologist without any reference to the hardness or compactness of the material to which the name is applied; thus he speaks of loose sand as a rock equally with a hard substance like granite.

Rocks are of two kinds, (1) those laid down mostly under water, (2) those due to the action of fire.

The first kind may be compared to sheets of paper one over the other. These sheets are called *beds*, and such beds are usually formed of sand (often containing pebbles), mud or clay, and limestone or mixtures of these materials. They are laid down as flat or nearly flat sheets, but may afterwards be tilted as the result of movement of the earth's crust, just as you may tilt sheets of paper, folding them into arches and troughs, by pressing them at either end. Again, we may find the tops of the folds so produced worn away as the result of the wearing action of rivers, glaciers, and sea-waves upon them, as you might cut off the tops of the folds of the paper with a pair of shears. This has happened with the ancient beds forming parts of the earth's crust, and we therefore often find them tilted, with the upper parts removed.

The other kinds of rocks are known as igneous rocks, which have been melted under the action of heat and become solid on cooling. When in the molten state they have been poured out at the surface as the lava of volcanoes, or have been forced into other rocks and cooled in the cracks and other places of weakness. Much material is also thrown out of volcanoes as volcanic ash and dust, and is piled up on the sides of the volcano. Such ashy material may be arranged in beds, so that it partakes to some extent of the qualities of the two great rock groups.

The relations of such beds are of great importance to geologists, for by means of these beds we can classify the rocks according to age. If we take two sheets of paper, and lay one on the top of the other on a table, the upper one has been laid down after the other. Similarly with two beds, the upper is also the newer, and the newer will remain on the top after earth-movements, save in very exceptional cases which need not be regarded here, and for general purposes we may look upon any bed or set of beds resting on any other in our own country as being the newer bed or set.

The movements which affect beds may occur at different times. One set of beds may be laid down flat, then thrown into folds by movement, the tops of the beds worn off, and another set of beds laid down upon the worn surface of the older beds, the edges of which will abut against the oldest of the new set of flatly deposited beds, which latter may in turn undergo disturbance and renewal of their upper portions.

Again, after the formation of the beds many changes may occur in them. They may become hardened, pebble-beds being changed into conglomerates, sands into sand-stones, muds and clays into mudstones and shales, soft deposits of lime into limestone, and loose volcanic ashes into exceedingly hard rocks. They may also become cracked, and the cracks are often very regular, running in two directions at right angles one to the other. Such cracks are known as *joints*, and the joints are very important in affecting the physical geography of a district. Then, as the result of great pressure applied sideways, the rocks may be so changed that they can be split into thin slabs, which usually, though not necessarily, split along planes standing at high angles to the horizontal. Rocks affected in this way are known as *slates*.

If we could flatten out all the beds of England, and arrange them one over the other and bore a shaft through them, we should see them on the sides of the shaft, the newest appearing at the top and the oldest at the bottom, as in the annexed table. Such a shaft would have a depth of between 10,000 and 20,000 feet. The strata beds are divided into three great groups called Primary or Palaeozoic, Secondary or Mesozoic, and Tertiary or Cainozoic, and the lowest of the Primary rocks are the oldest rocks of Britain, which form as it were the foundation stones on which the other rocks rest. These may be spoken of as the Pre-Cambrian rocks. The three great groups are divided into minor divisions known as systems. The names of these systems are arranged in order in the table, and the general characters of the rocks of each system are also stated.

With these preliminary remarks we may now proceed to a brief account of the geology of the county.

In considering the geology of Cheshire we may first glance at one of the coloured geological maps of England,

	Names of Systems	Subdivisions	Characters of Rocks
TERTIARY	**Recent Pleistocene**	Metal Age Deposits Neolithic ,, Palaeolithic ,, Glacial ,,	Superficial Deposits
	Pliocene	Cromer Series Weybourne Crag Chillesford and Norwich Crags Red and Walton Crags Coralline Crag	Sands chiefly
	Miocene	Absent from Britain	
	Eocene	Fluviomarine Beds of Hampshire Bagshot Beds London Clay Oldhaven Beds, Woolwich and Reading Thanet Sands [Groups	Clays and Sands chiefly
SECONDARY	**Cretaceous**	Chalk Upper Greensand and Gault Lower Greensand Weald Clay Hastings Sands	Chalk at top Sandstones, Mud and Clays below
	Jurassic	Purbeck Beds Portland Beds Kimmeridge Clay Corallian Beds Oxford Clay and Kellaways Rock Cornbrash Forest Marble Great Oolite with Stonesfield Slate Inferior Oolite Lias—Upper, Middle, and Lower	Shales, Sandstones and Oolitic Limestones
	Triassic	Rhaetic Keuper Marls Keuper Sandstone Upper Bunter Sandstone Bunter Pebble Beds Lower Bunter Sandstone	Red Sandstones and Marls, Gypsum and Salt
PRIMARY	**Permian**	Magnesian Limestone and Sandstone Marl Slate Lower Permian Sandstone	Red Sandstones and Magnesian Limestone
	Carboniferous	Coal Measures Millstone Grit Mountain Limestone Basal Carboniferous Rocks	Sandstones, Shales and Coals at top Sandstones in middle Limestone and Shales below
	Devonian	Upper Mid } Devonian and Old Red Sand- Lower stone	Red Sandstones, Shales, Slates and Limestones
	Silurian	Ludlow Beds Wenlock Beds Llandovery Beds	Sandstones, Shales and Thin Limestones
	Ordovician	Caradoc Beds Llandeilo Beds Arenig Beds	Shales, Slates, Sandstones and Thin Limestones
	Cambrian	Tremadoc Slates Lingula Flags Menevian Beds Harlech Grits and Llanberis Slates	Slates and Sandstones
	Pre-Cambrian	No definite classification yet made	Sandstones, Slates and Volcanic Rocks

DIAGRAM SECTION FROM SNOWDON TO HARWICH, ABOUT 200 MILES.

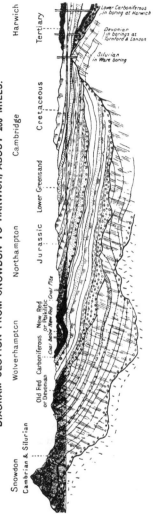

Snowdon — Cambrian & Silurian

Wolverhampton — Old Red Carboniferous / New Red or Poikilitic — Coal below New Red : Coal Pits

Northampton — Jurassic

Lower Greensand

Cambridge — Cretaceous

Harwich — Tertiary

Lower Carboniferous in boring at Harwich

Devonian in borings at Turnford & London

Silurian in Ware boring

The above section is intended to show the order of succession of the rocks in the crust of the earth as it would be seen in a deep cutting nearly E. and W. across England and Wales. It shows also how, in consequence of the folding of the strata and the cutting off of the uplifted parts, old rocks which should be tens of thousands of feet down are found in borings in the South East of England only 1000 feet or so below the surface.

where we shall see that almost the whole of Cheshire, a large portion of southern Lancashire, and parts of Staffordshire and Shropshire are coloured alike, usually in two shades of brown, for the Triassic rocks cover a large area in north-western England. The New Red Sandstone is the chief rock in Cheshire.

To the north, east, and south we shall see the coalfields of Lancashire, Yorkshire, and Staffordshire, and the limestone of Derbyshire, creeping near to the Cheshire boundary. But if we look carefully we shall notice that the eastern part of Cheshire is coloured differently from the rest of the county, and turning our eyes to the Welsh side of the Dee we shall find very ancient rocks rising in high hill-ranges.

The hills in the east of the county are much older than the sandstones of the plain, while the hills in Wales are ages older still, yet both are much higher than the main portion of Cheshire. We must therefore conclude that the surface of Cheshire has been at one time at a very different level, and that great physical forces have either depressed the centre of the county or have raised the hills surrounding it. An examination of the "strike" or "dip" of the rocks, the angle at which the strata are inclined, shows that the eastern hills have been twisted and contorted and slope downwards towards the west, and that the Welsh hills dip deeply eastward below the newer rocks of Cheshire.

The sandstones and other formations which stood upon the Millstone-grit and Carboniferous rocks have been worn away, and the older rocks left exposed. Again in the plain a similar process has taken place; rain and rivers have worn away the softer sandstone, forming hills and valleys, and the silt and wear of ages have left a fertile soil in most places, hiding the mother rock. We will begin by a description of these oldest rocks, and proceed step by step to the newer and later formed rocks.

The oldest rocks in Cheshire belong to the Upper Carboniferous series, and the Millstone-grit, a rough hard rock, overlies the Mountain Limestone, which is so noticeable in Derbyshire.

At one place, and one place only, Limestone comes near enough to the surface to be worth working. At Newbold Astbury, near Congleton, a cone of Limestone pushes up through the newer rocks and is quarried.

Above the Mountain Limestone are the Yoredale Rocks or Pendleside series, with a thickness of about 2000 feet, extending from Congleton Edge to Macclesfield, and eastward for about three miles. These Yoredale Rocks—shales and sandstones—extend for a short distance up the valley of the Goyt. A few fossils are found in Congleton Edge.

Above these again is the Millstone-grit, in which we find the Coal-measures. This is the most noticeable rock in Longdendale and southward along the Derbyshire border; it is estimated to vary from 1000 to 3000 feet in thickness. Geologists divide it into

> Rough Rock with Coal,
> Haslingden Flags with Coal,
> Roaches Grit,
> Kinder Scout Grit.

The Roaches are some rugged hills in Staffordshire, and Kinder Scout is the highest portion of the Derbyshire Peak. Roaches Grit is extensively quarried at Tegg's Nose near Macclesfield, and the freestone of Kerridge quarries, in the same neighbourhood, is largely used for roofing purposes. Kinder Scout Grit is a hard conglomerate containing many quartz pebbles; it weathers into fantastic shapes and edges. Laddow Rocks, in upper Longdendale, well exhibit the romantic weathering of this fourth grit.

The Coal-measures, which are mixed with the strata of the Grit, are exposed in a few places, but in most parts of Cheshire where coal is obtained the shafts are driven deeply through the New Red Sandstone to reach the Coal-measures. The Cheshire coal-fields are an outlying portion of the great Lancashire fields, but coal is worked so far south as Poynton near Macclesfield. Coal is also worked at Neston and Denna Collieries in Wirral, the levels running beneath the bed of the Dee; only what is known as the Lower Coal-measures are found in western Cheshire. Near Chester coal has been located at 510 feet below the Lower Keuper Sandstone. There are many fossils in the coal, and a few in the Millstone-grit.

Before we leave the Grit we may notice that it is the foundation of the extensive Cheshire grouse-moors. Heather grows well on the peaty soil, and in many places the rock is covered by thick mosses, which act as sponges to hold the water. From these great natural reservoirs the water trickles down into the rills, which growing into becks, and from becks into streams, provide Cheshire with pure and wholesome water, which is perhaps her most valuable commercial and agricultural asset.

Above the Coal-measures are the Permian rocks, followed by the Triassic, of which the most important are the Bunter and Keuper Sandstones. These rocks are noticeable in many parts of Cheshire, for instance at Alderley Edge, the Peckforton Hills, the Frodsham, Overton, and Delamere hills, and the outcrop near Runcorn. The Bunter lies beneath the Keuper Sandstone, and what is known as the Upper Mottled Bunter is extensively quarried at Runcorn, whilst at various places on the Frodsham hills both the Bunter and the freestone of the Lower Keuper are quarried for building purposes; the Bunter alone is in many places from 600 to 700 feet in thickness. Bunter which is quarried at some distance below the surface is uniform in texture and firm, but it is not useful for building stone, for it is unable to withstand the influence of the atmosphere and rots or " weathers."

The Lower Keuper is hard, and though in time it weathers it has more lasting qualities. The rugged tops of the hills at Halton, Weston, Overton, and Helsby are of this series; it is said that Beeston means "beetle stone," and that it derives its name from the rugged beetling rock. The conglomerate portion of the Keuper Sandstone is exceedingly hard, and is known to the quarrymen by the expressive name of " take-work," and the soft pure sandstone beneath is called " roach." Good hard sandstone and roach frequently lie in alternate layers, and the dark and light colours of the two qualities of stone show plainly in the sections cut in the quarries. Runcorn stone was

C. C.   3

Alderley Edge

used for the inside of the piers of the Menai Tubular Bridge.

Where the great forces of nature have cracked the rocks, throwing a formation out of its place, the crack or alteration in the strata, as has been already stated, is called a fault.   There are numerous faults in the sand-

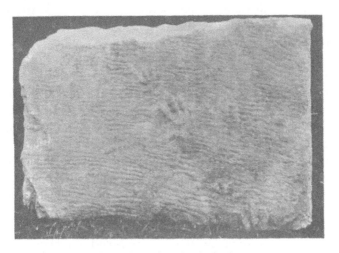

Footprint of Labyrinthodon

(*From the Storeton Quarries*)

stone, and in some places the Bunter actually stands at a higher level than the Keuper close to it.   Along the lines of many of these faults the softer rocks have been worn away and river valleys have been formed.

In the large and important quarries at Storeton in Wirral there are many estuarine deposits known locally as "footprint beds." In these beds, which are also found at Weston, Runcorn, and Daresbury, there are numerous ripple-marks, shales, "pseudomorphs" or altered crystals of rock salt, sun-cracks, worm-casts, and the footmarks and tracks of the creatures which walked about on the sandy shore before it had hardened into rock. The largest and most important of these footmarks, telling of days when the world was very young, are of gigantic batrachians, not unlike a newt in general form, known as *Labyrinthodon*, from the complex nature of their teeth. The footmarks vary much in size, and from fossil remains that have been found elsewhere it is certain that there were many species. At first, before any fossils had been found, these footmarks, very like the impression of a large hand, gave origin to the name of *Cheirotherium* or "hand-beast" for the unknown monster. These footmarks—which are preserved in many museums, and are even exposed on pavement flags in various parts of Cheshire—are a memorial of the wonderful creatures which hopped and walked about on the Cheshire beaches ages and ages before even the simplest form of mammal, of which Man is the highest in the chain of evolution, had appeared on the earth. Storeton quarries are everywhere famous for the abundance and perfectness of these geological records.

The top layer, if we can so express it, of the sandstone is the Upper Mottled Keuper, which, capped by boulder clay, is visible at the "Red Noses," at the extreme western point of Wirral, and at Burton Point. Off the coast of Cheshire, opposite West Kirby, runs a ridge of sandstone nearly two miles long, which in three places rises above the level of the highest tides and forms the three islands of Hilbre. On its wave-worn cliffs we can trace the various formations of the Upper sandstones.

Before we pass to a consideration of what lies above the sandstone, we will glance at the use of the rock itself. It yields three great sources of wealth and economy— stone, salt, and water. Mention has been made of the chief quarries, but there are many others in various places wherever the rock is not too deeply hidden by the clay and soil above it.

Rock-salt and brine springs occur in the lower beds of the Keuper Marl—the top layer of the New Red Sandstone—which in places is nearly 3000 feet in thickness. The Upper Keuper beds have been termed saliferous or salt-bearing strata. Rock-salt was first discovered at Marbury near Northwich in 1670, but salt had been obtained from brine springs by the Romans and perhaps earlier. The towns like Middlewich and Nantwich that have the termination "wich" or "wych" are salt towns. The layers of rock-salt vary in thickness; in the Marston mine at Northwich there are two beds, one nearly 90 feet thick and the other 150 feet, separated by about 30 feet of red clay. Gypsum is often found mixed with the salt.

In the New Red Sandstone proper—the lower Keuper and the Bunter—there is an abundance of clear fresh water, and for many years the towns in Cheshire drew most of their water supply from the rock. Even now, when water is brought from great distances in pipes, immense quantities are pumped up. Some 1,500,000 gallons per day are pumped from the Wallasey well, and the same quantity from the Flaybrick well at Bidston Hill by the Birkenhead Corporation. These wells are very deep, their borings going down in some places nearly 900 feet, passing right through the upper beds into the solid rock.

Copper-ore, cobalt, carbonate of lead, manganese, iron, vanadium, barium, sulphur, and other mineral constituents, are found in the sandstone, but seldom in sufficient quantities to be worth working. At Alderley Edge and Peckforton there were formerly extensive mining operations, copper being the metal mostly sought for, though cobalt and lead were also obtained. The Alderley mines will be mentioned in a later chapter. In a few places, for instance at Frodsham, Helsby, and Beeston, there are caves in the rock, which though ancient are not natural; they were originally formed by the digging out of the soft white lower sandstone for sanding the floors of cottages; the hard Keuper grit forms the roof of the cave. Soft stone from Storeton is ground for making concrete.

In the geological table of formations it will be seen that above the Triassic rocks come the Jurassic and Cretaceous Secondary rocks and all the Tertiary formations. It is not possible to say what once lay above the Cheshire sandstone, but the Lower Lias, which covers much of Shropshire, crosses the border in the south of the county, and is visible at Audlem in the form of shales and limestones. If, as has been suggested, Oolitic rocks covered the plain, they have been all worn away.

There are, however, deposits of great thickness above the New Red Sandstone in many parts, which entirely bury the rock beneath. This, as a whole, is known as the Drift, and consists of deposits laid down in geologically recent times, but only recent when compared with the ages of the more ancient rocks. Roughly, this is divided into Lower Drift sand, Lower Boulder Clay, Middle sands, Upper Boulder Clay, and Upper Drift sands.

Except in the Coal-measures few fossils are found in Cheshire, though we find traces of living creatures in the footmarks at Storeton and elsewhere, but in the Middle sands of the Drift teeth of at least two kinds of elephants, one of them being the mammoth, have been discovered in several places. As these sands were deposited by mighty glaciers, which have left the Cheshire clay full of what are called erratic boulders—that is masses of ice-carried rocks quite different in formation from the Cheshire rocks—it is quite possible that the elephant remains were also carried from a distance, perhaps from the mountains of Wales or the Lake District, from both of which glaciers found their way into Cheshire. These sands are used for building purposes, the clays for brick-making.

The Upper Boulder Clay often lies in flat terraces upon the sand, but near the hills the Drift trends upwards, and in some of the valleys amongst the hills is found to a considerable height. One granite boulder stands on a hill 950 feet above the sea. In the sands are found worn and rubbed marine shells, especially the common cockspur and the cockle, similar to shells now found living in the Irish Sea.

Over and above the Drift lie the alluvial deposits,

carried down by rivers or brought up the estuaries by the sea. These form the present soil of the land—gravels, sands, brick-earths and clays, and in many places peat. Remains of two kinds of extinct oxen are found in the alluvium, one being a wild and the other a domestic form, and bones of horses, goats, and other domestic animals are also present.

Beeston Bluff

From all the evidence of denudation and the deposit of drift and alluvium, we see that the plain of Cheshire has undergone great changes. It has been scooped and scraped by glaciers, leaving moraines, very evident in the Delamere country, and it has been washed by the sea or the waters of a great river. We can only suppose that, as Professor Ramsay suggested, it was at one time at an even lower level than it is to-day, and formed the bed of an important estuary, or of a river of much greater size than any we have in England now. It is clear that it was either an inland sea or salt lake, or an estuary, for the sandstone bluffs at Alderley, Beeston, and Helsby were once sea-coast cliffs, similar to but much grander than the Red Noses or Burton Point of the present day. Bowdon stands upon a great heap of estuarine sand, and in other places there is evidence of the wash of the sea along a now vanished coast-line.

We have so far only mentioned the name of peat, but this is a recent deposit of considerable interest and some commercial value which is cut, dried, and burnt for fuel. It was formed and is still forming on the hills and the few remaining "mosses" of Cheshire from the decomposition of the plants—chiefly a bog-moss—which grow in these wet places. The upper layers are light in colour and the lower ones very dark. On the Cheshire shore there are some ancient peat beds below the level of the high tides which are known as the Submerged Forest Beds; in these are numerous trunks of fallen trees, and in some cases the roots may be seen in their original position. In places they are covered by tidal sludge or blue clay. In the lower beds of the Wallasey shore remains of deer, cattle, dogs, and relics of an ancient barbarous race have been found, but in the upper beds there are coins and implements of Roman and also of mediaeval times, a large number of which are of thirteenth century date. We thus see that at no distant period this was dry forest land.

Buried Forest at Dove Point

The soil of Cheshire is fertile and the vegetable mould of the drained peat bogs is capable of high cultivation. The thick clay is firm and viscous, heavy to plough but productive of rich and plentiful grass for the farms. In almost every field in Cheshire there is a pond, now used for watering cattle, but originally dug for quite another purpose, namely to obtain clay or marl to spread over and enrich the fields. These ponds are called "pits," which is short for marl-pits, and were dug by regular gangs of "marlers," who went from farm to farm for that purpose. Nowadays artificial manures have destroyed the calling of the marlers. The American visitor who gets his first sight of English fields when he travels southward from Liverpool is always struck by the green of the Cheshire grass.

7. Natural History.

Various facts go to show that the British Isles have not existed as islands, separated from the Continent, for any great length of geological time. Around our coasts, for instance, are in places remains of forests now sunk beneath the sea, and only to be seen at low water. Between England and the Continent the sea is shallow, but a little to the west of Ireland we soon come to deep soundings. Great Britain and Ireland were once part of the Continent, and are examples of what geologists call recent continental islands. We have also no less certain proof that at some anterior period they were almost entirely submerged. The fauna and flora thus being destroyed, the land would have to be restocked with animals and plants from the Continent when union again took place, the influx coming from the east and south. Animals and plants, however, do not rush in a body into any new land, and the rate of distribution of various species varies, as also does the distance of their travels from their original homes. It was not long before separation occurred once more between Britain and the Continent, and only comparatively few of the continental species had spread so far west as Ireland before that island was cut off from England, and a few more reached England before it parted company with the mainland. Britain has fewer species than France or Belgium; the Isle of Man, Ireland, and the north of Scotland have fewer than the south of England. There are, however, other even more important factors which regulate the number of species in counties, and so far as these concern Cheshire we shall refer to them briefly.

The flora of Cheshire is small compared with that of Derbyshire and Flintshire, or with the southern counties of Kent and Surrey, but it is as rich and abundant as that of many southern counties. The comparatively short distance between Cheshire and the southern shires is not sufficient to influence the distribution of many plants. Where, however, the soil is either limestone or chalk we get a richer flora than on clay, and that is why Derbyshire and Kent have a larger number of species of plants on their lists than we have in Cheshire. In the west of England there are a few plants which established themselves when England was united with the Continent, but have failed to spread north and east away from the mild climate of Cornwall and Devon; these we do not get in Cheshire, but we do find on its hills an almost alpine flora which is absent from the south of England. The wild moorlands of Longdendale and the Peak are rich in berry-bearing plants and heaths, such as bilberry and cranberry, cowberry, the two commoner heathers and the ling; whilst the andromeda, bog-asphodel, and white cotton-grass abound on Cheshire moorlands. Cheshire lanes, even if their flowers fall in number of species below that of some other counties, are nevertheless ablaze with colour at different seasons. Hawthorn hedgerows, which border most of the lowland fields, are snowdrifts of blossom in spring, or pink with wild roses; waste places are golden with furze, the banks often so full of primroses that the violets hardly show, and even the spoil-banks in the salt-country are in early spring one sheet of glory with their abundance of that common weed the coltsfoot. Cheshire is, indeed, a garden of importance; the flora includes the grass of Parnassus and sundews of the hills, the buckbean and great spearwort of the meres, the sea-aster or wild Michaelmas daisy and seablite of the marshes, and the henbane and horned poppy of the shore. Lord de Tabley, the Cheshire botanist and poet, says :

> " And sheets of sapphire pave for miles
> The floors of hyacinthine woods."

This applies to most of the Cheshire woodlands, unless they are white with the beautiful but strongly smelling garlic. A volume would be needed to describe the Cheshire flora, but there is one point of special interest which should be mentioned. Cheshire has quite an important and special alien flora, plants which have been introduced unintentionally in merchandise and have firmly established themselves; we may mention in particular that near the Birkenhead Docks there are many American " ballast " plants, and that at Leasowe a curious composite, *Cotula*, has escaped from cultivation and spread all over the marshy land.

The admixture of mountain, plain, and sea-coast accounts for the large and varied fauna of Cheshire, but it is not possible to refer to more than a few of its animals. The mammals that are known to inhabit Cheshire exceed in number those of most other counties ; the large area of woodland—about one twenty-eighth part of the county—affording protection for rare and persecuted species. Nine out of twelve British bats have been recorded ; all the insectivores, and the badger, otter, stoat and weasel amongst the carnivores are plentiful, but the polecat and marten are practically extinct and there is no record of the wild cat. Thirteen rodents occur, though the dormouse and harvest mouse are rare ; the hare and rabbit abound on preserved land, and the mountain hare has been introduced and successfully established on the hills. In ancient days the royal forest of Delamere provided excellent sport for the Earls of Chester and the Kings of England ; James I was the last royal hunter of the wild red and fallow deer in Cheshire. Red deer, remains of the old stock, are now emparked at Tatton, Lyme, and Doddington. Herds of the so-called wild white cattle formerly existed in three Cheshire parks. At Lyme, where they roamed over the moorlands, they were finally destroyed in 1885, at Vale Royal they had ceased to exist much earlier ; but at Somerford their descendants are domesticated and milked.

Somerford White Cattle

Seals and whales occasionally reach our shores and are stranded on the sandbanks of Liverpool Bay and the estuaries. The common, grey, and hooded seal have been captured, and six different species of cetaceans have been stranded on the shore or banks. A grey seal was killed in the Mersey above Warrington in 1908.

The list of Cheshire birds is a long one, only inferior in numbers to those of counties which receive the great streams of migrants from the Continent or the birds of double passage which pass along our shores. Owing to its position, tucked away into one corner of Liverpool Bay, the Cheshire coast misses the main stream of western migration which passes up the Irish Sea from Wales to Scotland via the Isle of Man. Nevertheless many un-expected migratory wanderers have occurred, and the extensive banks and marshes in the estuaries attract immense numbers of waders, ducks, and wild geese, which either remain all winter or stop to feed on passage. The passerine birds which haunt the woodlands and open country are abundant, but the birds of prey are com-paratively few, the larger ones having been destroyed in the interests of game. The seafowl are plentiful in the estuaries but there are no breeding-places of rock-nesting sea-birds. There is, however, in Delamere Forest, one flourishing colony of black-headed gulls.

It is in the birds of the moorlands and meres that Cheshire is particularly strong. On the hills the red grouse reigns supreme; there are few better-stocked moors than those of the Yorkshire and Derbyshire borders. The black-game nest in the wooded valleys, the curlews amongst the heather. Nowhere south of Cheshire is the ring ousel so plentiful, and the dipper, grey wagtail, and common sandpiper abound on the moorland streams. The merlin, twite, and meadow pipit are birds of the heather, and the last-named frequently has the care of a young cuckoo, a bird which is seldom associated with moorlands but is really more plentiful on the hills than anywhere in the lowlands. In the lowlands the corncrake is perhaps more abundant than in any county further south, but the nightingale reaches its northern limit in Cheshire and is rare.

The great crested grebe is the most noticeable bird of the meres; in no other county, not even excepting Norfolk, is the bird so plentiful. The coot, moorhen, teal, mallard, tufted duck, and dabchick nest on or near the meres, and the reed warbler hangs its nest in the extensive reed-beds. The bittern frequently visits these reeds in winter, and perhaps might remain once more to nest if left unmolested, and great armies of migratory geese and ducks—mallard, teal, wigeon, tufted duck, pochard, golden-eye, shoveller, pintail, and goosander—come with more or less regularity in autumn and winter. Redshanks, ringed plovers, sanderlings, oyster-catchers, dunlins, knots, curlews, and whimbrels visit the estuaries on migration in countless thousands, but the bird of the estuaries which has increased enormously within recent years is the beautiful sheld-duck; almost yearly it extends its breeding range. Many rare wanderers reach the shores or meres on migration, amongst them the Kentish plover. One bird, Schlegel's petrel, has occurred in Cheshire and not only nowhere else in England but has never been observed elsewhere in Europe, or indeed in the Atlantic. A single example, a wanderer from the South Pacific, was picked up at Tarporley in 1908.

Reptiles are uncommon in Cheshire. Vipers are

almost extinct and ringed snakes are rare, the lizard and blindworm not abundant. The sand lizard occurs, or occurred until recently on the coast, its presence along a narrow strip of Lancashire or Cheshire sandhills being an exceedingly interesting exception to its general south-country distribution. Batrachians are abundant, the moisture of the county suiting their requirements ; the natterjack toad is not uncommon on the sandhills. The fishes of Cheshire will be mentioned in the chapter on Fisheries.

Molluscan life is varied though not so plentiful as in a limestone district, but insect life is abundant. There are, however, but few butterflies, though the night-flying moths are more numerous in species than in many parts. Blue butterflies are restricted to a few localities, and in one of these is found the Cheshire variety of the large heath butterfly. A peculiarity of the local lepidoptera is the abundance of the dark form of the peppered moth at the expense of the typical form. The list of Cheshire beetles, and indeed of all insects, is a long one, and this is especially noticeable of its scale-insects. The reason of this is that in Cheshire there have been a large number of hard-working naturalists, particularly amongst the working classes, who have devoted their well-earned leisure to the study of the fauna. It is not really that there are more insects than in other counties, but that more is known about them.

8.   Round   the   Coast.   Chester   to Warrington.

It is natural that we should start for an imaginary trip round the Cheshire coast from the county-town of

Water Tower, Chester; formerly on the bank of the Dee

Chester, and geographically Chester stands close to the junction of Cheshire and Wales. We have said that

4—2

Cheshire has a coast-line of over 40 miles, but as a matter of fact, though Chester stands on the tidal Dee, it is over eight miles from the coast. We shall have, therefore, to start our journey by land, for if we follow the course of the river, the artificial "cut" referred to in an earlier chapter will take us through Welsh ground. For a little over a mile the Dee Cop, the embankment of the river, runs through Cheshire, but the old river bank came almost under the walls of the city. All the way along the edge of Sealand we can see this old river bank, in most places now planted with trees. The county boundary, as a matter of fact, runs just below the bank, for it followed the river, and not its Cheshire bank.

At Blacon Point, now but a wooded bit of rising ground, we reach the first of the old headlands. There is an old Cheshire couplet which is supposed to describe the former abundance of trees in the forest of Wirral :

"From Blacon Point to Hilbree,
A squirrel could jump from tree to tree."

We must not forget that long before the Dee Cut was constructed Chester was an important seaport, and as we travel northward we shall see how by degrees the available landing-places or quays were pushed further and further from the city. The Dee estuary widens steadily; it is two miles across at Connah's Quay and over six between the Point of Air and Hoylake. It is choked with sandbanks and ever-growing marshes on the Cheshire side, and the Channel or Main Deep is for the greater part of its course on the Welsh or western side. Passing the village of Saughall, an unimportant place made famous in the seventeenth century by producing a horned woman, one of whose horns is said to be preserved in the Ashmolean Museum at Oxford, we reach the site of Shotwick Castle, which once guarded the approach to Chester. Nothing however remains to-day but a small mound, and Shotwick is mainly famous for its church, and its one-time vicar, Dr Samuel Clarke, author of the *Martyrology*. Twenty-five years ago the tide lapped the walls of Shotwick Churchyard, but now we look down on some of the most fertile fields in England.

Opposite Shotwick we may note a mound running out across the level, and we may perhaps have noticed the remains of two similar mounds nearer to Chester; these are all that is left of the banks or "cops" which reclaimed the marshes of Sealand. Beyond Shotwick we come to land which still retains its marshy character, where the redshank nests in spring and the grey goose feeds in winter. In front stands the tree-crowned sandstone headland of Burton Point, just beyond the embankment of the Great Central Railway, which has been more effective than any of the older banks in holding back the tide. This railway cuts right across the marsh from the swing-bridge at Shotton to Burton.

From Burton Point to the end of the Dee Cop runs the last of the retaining embankments, but in two places it was broken by storms soon after its construction and has never been mended. The tide passes through the gaps up to the strong railway embankment. The saltings north of Burton are in Cheshire, for here the boundary

line leaves the shore and runs down the centre of the
ever widening estuary. Marsh and slub, dotted at low-
tide with thousands of hardy Cheviot sheep, stretch for
miles, and beyond we can see the golden glimmer of the
"Sands o' Dee," immortalised in Kingsley's poem of the
tragic fate of Mary, which, we learn in *Alton Locke*,
was suggested by a picture by Copley Fielding and a
true story told to the Canon of Chester. Only those
who have seen the western wind "wild and dank with
foam" can realise how

> "The creeping tide came up along the sand,
>   And o'er and o'er the sand,
>   And round and round the sand,
>   As far as eye could see ;
> The blinding mist came down and hid the land—
>   And never home came she."

Mary, alas, is not the only unfortunate wanderer across
these trackless wastes who has failed to return ; almost
every year there are one or two added to the long list of
Dee tragedies.

Denhall, in its ilex grove, stands close to the water's
edge, and just beyond are the Denna Collieries which
were opened in 1750, and where at one time the coal
was carried through the narrow levels in underground
barges. The sandstone blocks, a heap of ruins, mark
the site of the quay where once ships of many tons burden
were moored to be loaded with coal.

We next come to the Old Quay House, marked on
old maps as "Old Kay," where two hundred years ago
there was a channel of upwards of thirty feet of water.

This was the first important quay after the port of Chester was given up; in the seventeenth century it was of far greater importance than any port in the Mersey. Shotwick Quay had been used in the fifteenth century as the port of embarkation to Ireland, but there was little oversea trade from the Dee until the Old Kay came into existence. The Quay House, still standing, has been used as a prison, a temporary barracks for soldiers waiting for the Irish packet, and as a private residence.

At Moorside was the wharf for Neston, and just beyond it the New Quay was constructed. This converted the river-side fishing-village of Parkgate into an important sea-port. For a time all the Irish packets and trading vessels for many ports sailed from Parkgate, but by the end of the seventeenth century Liverpool was becoming a busy port, and the difficulties of the navigation of the Dee drove trade to the Mersey. For a hundred years or more, however, Parkgate remained a fashionable sea-side resort, where even Londoners went as "sea-dippers." It is said that Handel, waiting for the Irish boat, composed the "Messiah" at the George Hotel in Parkgate, a story originated no doubt from his altering the score in the hotel before the oratorio was performed in Dublin. Parkgate almost fell into ruins when its fashionable days were past, but it has now become a pleasant residential district for Liverpool inhabitants.

Gayton lies between Parkgate and Heswall. In 1690 William of Orange stopped at Gayton Hall, after sailing from Chester to Gayton Cottage, once a quay-side inn, now a residence, before he crossed from Hoylake to

Ireland and fought the Battle of the Boyne. Above Heswall is the Beacon Hill, but the name is probably of older origin than Dee navigation; the beacon was doubtless lighted here by Saxon defenders to warn the country side of the approach of Danish pirates. The light may have "flared to heaven" when the message of the approaching Armada was borne by fiery sign from hill to hill, for a few years later Heswall men, amongst many others, were instructed to keep their beacon in order and to post reliable watchers.

Thurstaston derives its name from a large rock on the breezy common above the village. "Thor's stone," tradition says, was placed there by the Danes to commemorate the great battle in Wirral in which they were defeated, a rather unlikely explanation. Telford, the great road engineer, constructed or partly constructed, in 1822, the now vanished port of Dawpool, really the last of Chester's outlying quays. Dean Swift disembarked here on his return from Ireland in 1707, and rode to Parkgate, but there is no certain record of "Dorpoll" as a busy port, for Telford's later attempt to convert it into the point of embarkation for Ireland failed through the rapid shoaling of the channel.

West Kirby, never a port, is now a thriving residential town and watering place; it is said that Christian missionaries from Ireland first landed here in Cheshire, and founded one of their earliest mission churches in England. Opposite West Kirby are the islands of Hilbre, already mentioned in the chapter on geology, where there is a signal station of the Mersey Dock and Harbour Board.

Rounding the Red Noses we face the Irish Sea and are in Hoylake. The Royal Hotel, then a residence of Sir John Stanley, and a few fishing huts, formed Hoylake in 1792; now it is the sister-town of West Kirby. Hyle Lake, as it was called, where large ships could a hundred years ago lie in safe anchorage, is now silted up by the southward extension of the East Hoyle Bank, the largest sandbank in Liverpool Bay. The monks of St Hilburgh had a chapel on Hilbre Island in the eleventh century, which they reached by skirting the southward margin of the lake at low water. Hilbre, much later, was a busy spot, occupied by people of doubtful character who sold drink to and in other ways lived upon the sailors whose ships were lying at anchor in Hyle Lake. The great Hoyle Bank is famous for its cockles, and part of the population of Hoylake is engaged in the industry of cockling.

The land from Hoylake to New Brighton, at the mouth of the Mersey, was not many years ago covered with sand dunes, but the construction of the Leasowe embankment and extension of the Hoylake promenade have made it possible to convert these shifting dunes into building sites or to put them to other uses, for instance the formation of golf-links. Following the shore we come to Dove Point, where at low tide we may see the timber of the old Wirral forest embedded in the submarine peat. We pass the now forsaken Leasowe lighthouse, and soon reach Leasowe Castle, which was built about 1593 by an Earl of Derby. From his family it passed into the hands of the Egertons, and during the Civil Wars fell into ruin and earned the name of "Mockbeggar Hall";

New Brighton

the sands on our left are still known as Mockbeggar
Wharf. The Hall was subsequently rebuilt and occupied
by the Custs, and now its fortunes have changed once
more and it has become an hotel.

The last stretch of sandhills extends from Leasowe
Castle to New Brighton, a busy pleasure-resort of Liver-
pool and Lancashire workers. New Brighton lighthouse
and fort stand at the western entrance to the Mersey.
The next six miles of Cheshire coast is practically
Birkenhead and its outskirts. As we travel southward
up the estuary we pass in quick succession the various
piers of the river ferries—New Brighton, Egremont,
Seacombe, Woodside, Tranmere, Rock Ferry, and New
Ferry, to and from which the cleverly-handled ferry-
boats pass carrying passengers between Birkenhead and
Liverpool. Nowhere in England is there a busier ferry
traffic than across the Mersey.

Between Seacombe and Woodside is the entrance to
the Great Float, once an inlet into which the tiny Birket
flowed, now an inland sea, lined with docks and wharfage.
Birkenhead, now nicknamed "Liverpool's Bedroom," was
but a tiny village clustered round an ancient ferry little
over a century ago, but now the forest of masts which
line the Cheshire shore, and the network of tramlines
converging on the bustling Woodside Ferry and Birkenhead Station, at once point to it being the town of
greatest importance in Cheshire. Chester is but a quiet
country town compared with Birkenhead. As part of
Greater Liverpool it is one of the principal centres of the
world's commerce.

Tranmere, Rock Ferry, and New Ferry are suburbs of Birkenhead, and beyond them we come to Lower Bebington, near the entrance to Bromborough Pool. Port Sunlight, a model village for the workers at Lever Bros. soap-works, stands at the mouth of the pool, on land which twenty years ago was marsh and nothing else.

Ferry Boats on the Mersey

The houses are in various styles but many of them are reproductions of the old Cheshire black and white.

Bromborough is the next village of importance, and here on the ordnance map we shall find two names in Old English characters, which means that they are of historical importance. The first of these is St Patrick's

Well, for we are told that St Patrick sailed from Ireland and entered the Mersey, and for a safe landing-place chose Bromborough Pool—where he found and blessed the spring that bears his name to-day. The other name is "Wargraves," and is given to two fields between the road and the river. Without doubt this marks the site of a battle, but what the battle was has never been clearly proved. In the year 937 Aethelstan and Edmund Aetheling led the Wessex and Mercian soldiers against a large army of invading Danes, who had come, it is said in the Chronicles, in 615 ships. The English met and defeated this army, which consisted of some 30,000 men of various nationalities serving under the Danes, at a point close to where they had disembarked known as Brunanburgh. Yorkshire claims this battle as having been fought on the Humber, Cheshire upholds the theory that Brunanburgh was but the ancient name of Bromborough, which even three hundred years ago was spelt Brunburgh. At any rate, Anlaff, the Danish leader, self-constituted king of Northumbria, had with him Scotch, Irish, and Welsh princes and their troops; so it is possible that this great battle was fought on the Wirral plains.

A mile south of Wargraves is Eastham pier, which was for many years the point of embarkation for Liverpool. A coach ran from Chester in connexion with the sailing ferry-boats, which often took many hours to do the five-mile journey. A few yards south of the pier are the tidal locks at the Mersey end of the Manchester Ship Canal, which from here to Warrington runs along the

Cheshire side of the river, thus avoiding the banks and shifting sands of the estuary. Nathaniel Hawthorne described Eastham as the finest old English village he had seen.

At Liverpool the Mersey is scarcely three-quarters of a mile across, but a little to the south of Eastham it has widened out to three miles or more—a great inland sea at high water, a waste of banks intersected by narrow and difficult channels when the tide is out. It has been suggested that this once was an inland sea, the "mere sea," but there is no geological evidence to support this theory, and probably the wastage of the banks, caused by the strong tides washing in through the narrow rocky neck, accounts for its present shape and size, which now cannot be enlarged on the Cheshire side because of the strong embankment of the Ship Canal.

As we travel along the canal bank we pass two spots of historic interest. First is the present racing and sporting centre of Hooton Park, which once was a seat of the Earls of Derby, but the old home of the Stanleys was replaced by a more modern building at the end of the eighteenth century. The next old building, Poole Hall, is of much earlier date and part of the sixteenth century walls remain to this day; it too was the home of an old Cheshire family, the Pooles, who resided here from the days of Edward I until but a few years ago. The Ship Canal is widened to 120 feet as we approach Ellesmere Port, and near Poole Hall is a great mound, quite a substantial hill, known as Mount Manisty, now the home of rabbits and sheld-ducks. It is an artificial hill, a gigantic spoil-bank from the Canal, named after the engineer of the section.

Ellesmere Port and Whitby Locks were here before the Ship Canal, but their importance has grown since its construction; they were the river end of the Ellesmere Canal, which started at Ellesmere in Shropshire, and is now known as the Shropshire Union Canal. The place has grown by leaps and bounds since the Ship Canal came, and now there is a little forest of masts of coasters, and a fleet of flats moored always alongside the great elevator and the wharves.

A little rocky headland, Stanlow Point, is now cut off from the land by the Canal; on it to-day are a farm and a cottage or two, but once it was a secure islet standing amongst the almost impassable marshes, on which stood the Cistercian monastery of Stanlow. There are still a few traces of the stones and underground passages of the Abbey which was built in 1178, but only the eye of an expert can detect them. The insecurity of their wave-washed habitation was evident to the monks soon after they had settled in this out-of-the-world spot, but they remained here for over a hundred years, until their land was eaten away by the sea, one of their towers fell, and fire destroyed a portion of the building. Most of the monks were then removed to Whalley in Lancashire, but Stanlow remained as a cell of Whalley Abbey until the Dissolution.

Ince, with its Hall, and, in the village, the remains of one of the granges of Stanlow still standing, is the next place we pass after crossing the spot where the

Gowy is syphoned under the Canal. Then we travel between the Ince and Frodsham marshes, and the narrow rim of saltings on the estuary side of the Canal known as Frodsham Score, until we reach the busy entrance of the river Weaver and the Weaver Canal at Weston Point. Above us are the quarries on the face of Beacon Hill, before us, until we reach Runcorn, the wharfage and docks of the Ship Canal.

At Runcorn Gap, where the river narrows suddenly, only to widen out again above, there is a fine transporter bridge connecting Runcorn with Widnes. Here, though the river is tidal to Warrington, we practically leave the coast.

9. The Coast—Gains and Losses. Pro= tection of the Coast.

The coasts of an island are liable to changes which are regulated by the nature of the rock or soil forming the shore and the set of tides and currents which wash it. The Cheshire coast is not an exception to the general rule, and during the last few hundred years it has altered. The most noticeable gains have been in the upper part of the estuary of the Dee, the greatest losses in the Mersey. It will be remembered, from what we have stated in the last chapter, that the Dee formerly flowed much to the north-east of its present channel through Welsh land, that indeed it passed right along the Cheshire side of the reclaimed land at Sealand from

The Transporter Bridge, Runcorn

Chester to Burton. Reclamation of Cheshire land began, however, before the construction of the Dee Cut. In the fifteenth century the navigation of the river to Chester was difficult owing to the growth of the sandbanks, but it was not until the middle of the eighteenth century that Sir John Glynne reclaimed, by means of an embankment, a large portion of what was up to that time saltings and sand on the Cheshire side of Saltney. By 1737 the New Cut was practically complete, and "cops" had been run across the flats of the estuary on the Welsh side of the old river; in 1790 the Shotwick Cop was finished and large tracts of land were brought under cultivation. About 1860 the Burton Cop was made, and from that day marsh-land, suitable for the grazing of sheep, which some day will be finally enclosed, has been rapidly forming along the Cheshire shore from Burton towards Parkgate. Great accumulations of sand have already formed opposite Parkgate and Heswall, and, though it takes time for marsh and grass to become established on sand, the end is certain. Slowly but surely the influence of the Burton Cop and the training-wall, which runs towards Flint as a continuation of the Chester Cop, will convert the sands of Dee into dry land, and Cheshire will gain a huge tract of useful country, similar in nature to the fertile plain of Sealand. A sure sign of the change is the constant alteration of the moorings of the Parkgate fleet of shrimpers. A few years ago they were anchored opposite the town, then they were moved to Gayton, and now they can barely find water in the gutter at Heswall.

North of Heswall the action of the tide has worn away the marl cliffs, and certainly much ground has in former years been lost, but the advance of the marshes will probably result in this land being eventually regained.

In Liverpool Bay there is certain evidence of submergence in the buried forest at Dove Point, but so constant are the changes in the shape and size of the sandbanks off the coast that it is impossible to estimate the former size of this lost land. The set of the tide into Liverpool Bay brings back the silt from the rivers and deposits it on the banks, which must be very many feet in thickness, and some of them, as for instance the East Hoyle Bank, have grown so rapidly as almost to destroy the anchorage known as Hoyle Lake and similar deep-water roads. Only by constant dredging is the channel into Liverpool kept clear.

Probably at one time the reef of rock on which stand the islands of Hilbre was in direct connexion with the Red Noses at Hoylake, but not in historic times; the tendency now is for the two to become connected once more by sand, though not by rock.

The formation of the Leasowe embankment, as already stated, has prevented further sea incursions on the north coast of Wirral, and in the same way the Ship Canal embankment has saved the land on the south bank of the Mersey estuary. The historical evidence of the complaints of the monks of Stanlow shows that in the twelfth century the sea was constantly threatening to cover more and more land.

10. The Coast. Sandbanks and Lighthouses.

Anyone who travels on one of our great ocean liners out of Liverpool may notice that the course taken is northward for six miles before the steamer turns westward. The reason of this is that the only really deep-water channel runs along the Lancashire coast, and that on either side of it are shallow banks of sand. The first of these seaward is the Great Burbo Bank, and the others all have names well known to the Liverpool pilots. There is, however, another deep-water channel, known as the Rock Channel, which turns sharply westward at the Rock Lighthouse at New Brighton. It is too narrow and difficult to navigate for a large steamer, but is used at high water by all small craft which are steaming westward. Almost opposite Hoylake this channel turns northward, and runs, as the Horse Channel, along the eastern edge of the Hoyle Bank.

The navigation of the approaches to Liverpool is difficult, and even the experienced Liverpool pilots require some marks to guide them both by day and night. All round our coasts are lighthouses, lightships, beacons, and buoys, placed in certain positions for the purpose of guiding mariners. Very little, however, was done to help sailors in this way until the beginning of the nineteenth century, although the Romans set a good example some 2000 years ago by the erection of beacons and towers, some of which are still standing to this day.

Rock Lighthouse, New Brighton

The corporation of Trinity House, an "association for piloting ships," was founded in 1512, and at the present day the Elder Brethren of Trinity House are responsible for the erection and maintaining of the lights and marks on our coast which regulate the movements of shipping. Their income is derived from light-dues levied on shipping, and amounts to about £300,000 a year. There were about 30 lighthouses round the British coasts a hundred years ago, but now there are nearly 900.

We have already mentioned the beacon on Heswall Hill, and if we look carefully at an old map of Cheshire we shall find quite a number of "beacon hills." As some of these are a long way from the coast it is quite evident that they were not all used as guiding lights for vessels, and we know that, until the electric telegraph came into use, beacon fires were the only means of sending urgent messages quickly through the land. We have read how the approach of the Spanish Armada was signalled by these fires, and how quickly the whole of Britain was roused. There were beacons at Heswall, Caldy, Bidston, and Weston Point, but the two first were certainly originally placed as warning fires, probably in Saxon days when there was so much danger from the piratical Danes and Irish.

Very much later these beacon towers were used as marks by day by sailors in Liverpool Bay, and on Grange Hill, a little way from the beacon, a windmill served this purpose, and was replaced by a permanent tower when it was blown down in a storm.

Until quite recently the approaches to Liverpool, Birkenhead, and Chester were marked by lighthouses at the Point of Air on the Flintshire side of the Dee, at Hoylake, Leasowe, Bidston, and New Brighton in Cheshire, and at Formby and Crosby in Lancashire. Bidston and New Brighton remain, but the other lighthouses have been replaced by lightships and by buoys, of which many are illuminated. The reason of this is that the channels in Liverpool Bay are often altered by the shifting of the banks, and fixed lights are after a time a danger rather than a guide.

It is said that the earliest light in Cheshire was erected by the African slave-traders of Liverpool on the shore at Meols. This gave place to the Leasowe light, taken over by the Dock and Harbour Board, but before this lighthouse was finally given up, its position had several times to be altered; indeed about 1771 it was washed away. The present tower, which is 40 feet high, was built in 1865. In 1771 a lighthouse was built on Bidston Hill, and about the same time two, a high and a low tower, were erected at Hoylake. All three lighthouses were re-built later.

Bidston light now stands 225 feet above the sea, and has a fixed white light visible 23 miles at sea; it is masked or invisible east of a line from the Bar Lightship. After vessels coming into the river fail to see this light they know they must alter their course. The lighthouse at New Brighton, which is built of Anglesey granite, was put up in 1829; every 20 seconds it flashes a white light, which is visible 14 miles. It also shows both by day and night a signal which means that there is or is not water enough in the Rock Channel.

Liverpool Bay and the Mersey Channels are now lighted by the North-west, Bar, Formby, and Crosby lightships, and the entrance to the Dee by the Dee lightship. The Bar light, about 12 miles by the channel from New Brighton, has a triple white flash every 30 seconds which is visible 10 miles away. Every 20 seconds the Formby lightship flashes a red light, and every 10 seconds the Crosby ship a white light; the mariner picks one up after the other as he comes into the river.

The main channels are marked with buoys, red conical buoys on the starboard hand of a vessel entering the river, and black "can" buoys on the port. Besides these are buoys shaped or coloured in different ways to show particular steering points. More than 36 of these buoys are lighted by compressed gas, the starboard buoys showing white, the port red flashes. These flashes are regulated by clockwork, and each buoy carries enough compressed oil-gas to last for several weeks without refilling. Opposite New Brighton, the point where the rapid tideway enters the river is known as the "Rip-Rap." This rough channel is marked by an experimental buoy lighted by a flashlight of acetylene gas of 340 candle-power which stands 10 feet above the sea. The gas generator carries 2000 lbs. of carbide, and it has already been proved that the single charge will last without any attention for six months. The light is so powerful that mariners say they can pick up this light when they cannot see the oil-gas buoys close to them, and the experiment has proved so successful and economical that other buoys are being converted to this system. Eventually, in all probability, all the buoys in the Bay will be converted, and then Liverpool Bay will be the best-lighted harbour approach in the world.

It will have been noticed that the only lights actually on the Cheshire coast are at Bidston and New Brighton, but one or two of the lighted or beacon-buoys are so placed that they serve as shore-lights. Examples of these are the fixed red light known as the Meols Outfall Beacon, and the red-flash buoy at Hilbre.

In the Mersey all the dock entrances and the piers are lighted with distinctive lights, and up the river there is a lightship off Garston, a lighthouse at Hale Point in Lancashire, and one at Weston Point at the mouth of the Weaver.

It has often been suggested that the fortifications of the Mersey are antiquated and useless, and that an invading navy could easily subdue them. The natural protection of Liverpool and Birkenhead is, however, of great importance, and it would only be necessary to extinguish the lights to make it difficult for even a Liverpool pilot to enter at night, while if the buoys were moved a little from their present position the navigation of Liverpool Bay would be almost impossible.

11. Climate and Rainfall.

The climate of a country or district is, briefly, the average weather of that country or district, and it depends upon various factors, all mutually interacting, upon the latitude, the temperature, the direction and strength of the winds, the rainfall, the character of the soil, and the proximity of the district to the sea.

The differences in the climates of the world depend mainly upon latitude, but a scarcely less important factor is this proximity to the sea. Along any great climatic zone there will be found variations in proportion to this proximity, the extremes being "continental" climates in the centres of continents far from the oceans, and "insular" climates in small tracts surrounded by sea. Continental climates show great differences in seasonal temperatures, the winters tending to be unusually cold and the summers unusually warm, while the climate of insular tracts is characterised by equableness and also by greater dampness. Great Britain possesses, by reason of its position, a temperate insular climate, but its average annual temperature is much higher than could be expected from its latitude. The prevalent south-westerly winds cause a drift of the surface-waters of the Atlantic towards our shores, and this warm-water current, which we know as the Gulf Stream, is the chief cause of the mildness of our winters.

Most of our weather comes to us from the Atlantic. It would be impossible here within the limits of a short chapter to discuss fully the causes which affect or control weather changes. It must suffice to say that the conditions are in the main either cyclonic or anticyclonic, which terms may be best explained, perhaps, by comparing the air currents to a stream of water. In a stream a chain of eddies may often be seen fringing the more steadily-moving central water. Regarding the general north-easterly moving air from the Atlantic as such a stream, a chain of eddies may be developed in a belt parallel with its general direction. This belt of eddies, or cyclones as they are termed, tends to shift its position, sometimes passing over our islands, sometimes to the north or south of them, and it is to this shifting that most of our weather changes are due. Cyclonic conditions are associated with a greater or less amount of atmospheric disturbance ; anticyclonic with calms.

The prevalent Atlantic winds largely affect our island in another way, namely in its rainfall. The air, heavily laden with moisture from its passage over the ocean, meets with elevated land-tracts directly it reaches our shores—the moorland of Devon and Cornwall, the Welsh mountains, or the fells of Cumberland and Westmorland—and blowing up the rising land-surface, parts with this moisture as rain. To how great an extent this occurs is best seen by reference to the accompanying map of the annual rainfall of England, where it will at once be noticed that the heaviest fall is in the west, and that it decreases with remarkable regularity until the least fall is reached on our eastern shores. Thus in 1906, the maximum rainfall for the year occurred at Glaslyn in the Snowdon district, where 205 inches of rain fell ; and the

lowest was at Boyton in Suffolk, with a record of just
under 20 inches. These western highlands, therefore,
may not inaptly be compared to an umbrella, sheltering
the country further eastward from the rain.

The above causes, then, are those mainly concerned
in influencing the weather, but there are other and more
local factors which often affect greatly the climate of a
place, such, for example, as configuration, position, and
soil. The shelter of a range of hills, a southern aspect,
a sandy soil, will thus produce conditions which may
differ greatly from those of a place—perhaps at no great
distance—situated on a wind-swept northern slope with
a cold clay soil.

In Cheshire, which at one extremity touches the sea,
and at the other has a considerable elevation, there is
very marked difference in the climate and rainfall according to locality. We shall find on glancing at a map
coloured to show annual temperature in England, that
western Cheshire is much warmer than eastern Cheshire.
In the west it is on the average the same as in the
Midlands and even in Norfolk and Suffolk, but in the
east it is more the temperature of northern England. The
explanation of this difference is that the eastern part of
Cheshire is high land, part of the great hill-country, where
the rainfall is considerable and the soil is spongy moss on
a rock foundation. In the west we have the clay soils,
with a more even temperature, the neighbourhood of the
sea, and a much greater area of land in which woodlands
are plentiful; all of which factors are in favour of a mild
climate.

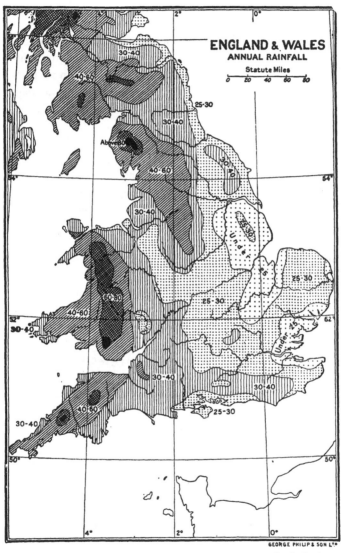

ENGLAND & WALES
ANNUAL RAINFALL

Statute Miles
0   20   40   60   80

(*The figures give the approximate annual rainfall in inches.*)

The Royal Meteorological Society receives reports of the temperature of the air, the hours of sunshine, the rainfall, and the direction and force of the wind from all parts of our islands, and publishes a daily report with a map, which may be seen in almost any newspaper. We can at a glance learn what has happened in other parts of England during the last twenty-four hours. At the end of the year these results are all brought together and averaged, and we can compare this average climate of the different British areas. Let us look at the climate in 1907, as it is recorded for Cheshire as compared with other places.

In 1907 the mean temperature of England was $52 \cdot 6°$, while that of Cheshire was $48 \cdot 2°$, varying from $49 \cdot 0°$ at Hoylake to $46 \cdot 8°$ at Macclesfield. Thus we see that Cheshire as a whole is below the average for the whole of England.

Bright sunshine was recorded at Greenwich Observatory in 1907 on $1405 \cdot 8$ hours, and at Hoylake in Cheshire there were $1477 \cdot 3$ hours. If, however, we were to compare these figures with those for the east, where there is more accumulation of moisture on the hills, we should find more hours in the west than in the east. In Cheshire, as a whole, there were 58 sunless days.

Now turn to a map shaded to show the variation in rainfall, which is of course a very important factor with regard to the commercial and agricultural welfare of a district. The annual rainfall, it will be seen, is greatest in the mountains of the Lake District and Wales, and least in the eastern counties. In 1907 the highest rainfall, where 196·16 inches were measured, was at Llyn Llydaw, Snowdon, in Carnarvonshire, and the lowest, recorded as 16·66 inches, at Clacton-on-Sea, in Essex. The average for the whole of England was 33·61 inches, and Cheshire was a little above the average with a rainfall of 34·93 inches. Looking at the map we shall see that this average is obtained in Cheshire in rather a curious way, for there is a tongue of light colour, meaning light rainfall, running up from the Midlands and eastern counties in a north-westerly direction, and a dark tract, terminating in the Peak of Derbyshire, which comes down from the north along the centre of England and covers a portion of eastern Cheshire. Comparing this map with one that shows the levels, it will be seen that the heavy rainfall is over the grit hills, and the light fall over the level plain.

If we take the average of a number of years, we shall see that on the hills the highest rainfall is between 40 and 60 inches; in the western part of the lowlands is a light rainfall of below 30 inches, and in Wirral and a tract of land lying between the two extremes it is between 30 and 40.

The highest rainfall in Cheshire in 1907 was at Swineshaw, near Stalybridge, where 53·20 inches were recorded, and the lightest at Thurstaston, in Wirral, where only 22·90 inches fell. There were 203 "rain days" (that is days when more than ·010 inch was measured) in England in 1907, and 217 rain days in Cheshire. At Northwich there were 222, and at Hoylake 208 rain days.

Because Cheshire is so well watered, and on account

of its somewhat heavy clay soil, fogs are of rather frequent occurrence in the autumn and winter. Sea fogs are occasionally very troublesome in the Mersey, where they are thicker than in the Dee, but perhaps the worst fogs are in the north-east of the county, for the thick smoke of Manchester and Stockport, and other manufacturing towns in or near the borders of the county, is kept down by the heavy atmosphere, and renders the fogs exceedingly unpleasant and dirty.

The prevailing winds are south and south-west. The eastern portion of Cheshire is sheltered from the bitter east winds by the high land of the Pennines. A north-west gale in winter drives up the Dee estuary, and converts what is usually a peaceful shallow-water arm of the sea into a storm-swept dangerous place ; the full force of heavy seas is not, however, so severely felt on the Cheshire shore as it is on the sandbanks of Liverpool Bay, which act as a natural breakwater for the coast.

12. People—Race, Dialect, Population.

Very little is known about the Cornavii, the British tribe which before the Romans came inhabited what is now Cheshire. Their origin, their customs, even the boundaries of their territory, are only matters for guess-work. That they had a town of some importance where Chester stands to-day on the Dee is certain, and from the proximity of Chester to the secure retreats in the Welsh hills it is probable that the influence of intercourse with the ancient Keltic races was felt in Cheshire.

Cheshire was civilised by the Romans, and long after the victors had to leave our shores Romano-British industries were carried on at Chester, on the Mersey at Stockton Heath, and to some extent in the salt country at Northwich. The Keltic tribes speedily took possession of Chester when the Romans had left it, and though the English invaders succeeded in reaching it from Wessex they made no prolonged stay until early in the seventh century when the Bernicians, afterwards Northumbrians, took the city. Once more the Welsh got the upper hand, but in 828, when the territory became part of Mercia, it was peopled by Saxons.

The language now spoken in England is based on the speech of our Saxon ancestors, and in Cheshire the people being under the influence of the northern men spoke a northern dialect. When the Normans conquered the Saxons, dialects increased, but that of Cheshire remained one of the northern forms, in which there is a tendency to shorten the vowels, and to clip words. There is little pure Cheshire spoken to-day, for in the north and east the influence of the Lancashire dialect is very marked; in the south of the county the folk-speech is very similar to that of Staffordshire and Shropshire, and about Chester it is not unusual to hear Welsh words in ordinary conversation.

There is one rather remarkable exception to the rule that near a county boundary the dialect is influenced by that of the neighbouring county, and it is no doubt explained by the greater isolation of a district when its bounds are hills than when they are rivers. In the valley of Longdendale many of the local names for objects are entirely different from those in use in Lancashire and Yorkshire only a few miles to the north.

There were no settlements of foreigners in Cheshire sufficiently important to leave much mark upon its commerce. The woollen trade of East Cheshire, perhaps introduced by Flemings or Walloons, was swept away by the spread of the cotton trade, and the silk industry of Macclesfield was founded by local enterprise and not by alien settlers. Cheshire has long remained in the hands of Cheshire men, and many Norman and a few Saxon names still exist in all parts of the county. Speed, who was proud of his Cheshire origin, said that the men of Cheshire, "for Nature's endowments (besides their noblenesse of mindes), may compare with any other nation in the world," and William Smith, at the end of the sixteenth century, describes them as "stout, bold, and hardy; of stature tall and mighty; withall impatient of wrong, and ready to resist the Enemy or Stranger, that shall invade their Countrey," and the women he calls "very friendly and loving."

The population of Cheshire has increased in certain districts since the census was last taken, and over fourfold since 1801. In that year the population was 192,305, and in 1901 it was 827,191. The number of people to the square mile in England as a whole in 1901 was 558 and in Cheshire 808. Birkenhead, Stockport, Hyde, Stalybridge, Dukinfield, and Macclesfield are the most thickly populated areas, and in these towns and the districts lying immediately to the south of Manchester the increase is most marked.   In the country districts and in some of the older towns there has been a slight decrease since the census of 1891.

In 1901 there were 395,017 males and 432,174 females in Cheshire.   People of foreign origin, largely from the east and engaged in the Manchester trade, numbered 4147; there is a large foreign population in the port of Birkenhead.   There were, when the census was taken, 169,123 inhabited houses, and 2102 people on shipboard, canal barges, and boats.   In military barracks there were only 248 officers and men.

In the chapters dealing with the trade of Cheshire the numbers engaged in various occupations will be found, but here we may mention that, unable or unwilling to work in ordinary ways, there were 8810 persons in hospitals, workhouses, asylums, and industrial schools.

13.   Agriculture — Main   Cultivations, Woodlands, Stock.

We stated very early that Cheshire is an agricultural county, but that does not mean that it is a great corn-growing area.   If we look at the report published every year by the Board of Agriculture, in which the following divisions are arranged—-corn crops, green crops, clover, saintfoin and grasses for hay, grass not for hay, flax, hops, small fruit, and the numbers of various kinds of live stock— we shall realise the nature of the agricultural products of Cheshire and their value.   It is very interesting to note that out of a total area of nearly 648,000 acres, only 226 are fallow, which means that nothing is being produced from them. It is, however, necessary to state that this does not include the grouse-moors of the east, 14,000 acres, which are used for grazing.

Corn crops are grown on 82,905 acres, or about one-eighth of the county; oats, largely used for its straw, is the most important corn crop, the acreage under this cereal being 64,953 against 13,713 acres upon which wheat is grown. Hardly any peas or beans are grown, and the crops of rye and barley are not considerable.

Amongst the green crops we find that 23,084 acres are used for growing potatoes, a very considerable area. 15,600 acres are devoted to turnips, swedes, mangolds, and cabbage, much of these being used as winter food for cattle. Other green crops are insignificant, and hops are not raised at all. Some 326,805 acres are under permanent pasture for cattle, this being a little more than half the total area. About another eighth of the county is used for the growing of clover and grasses, the land being broken up in rotation; of this 64,000 acres is for hay. Of the permanent grass rather less than one quarter is for hay, making a total of 156,414 acres for hay.

The growing of fruit in Cheshire has within late years received considerable attention, and now small fruits are grown for the markets of various large towns. Some 713 acres, mostly in the Dee Valley near Farndon, are devoted entirely to strawberries, and about 800 to mixed crops of strawberries, raspberries, currants, and gooseberries. Apples, pears, and stone-fruits are grown on nearly 2000 acres, but the return of only 56 acres for plums gives no idea of the quantity of damsons which

Marton Oak, one of the oldest Trees in Cheshire

are grown in the county. The damson crop is often a very large one, and Cheshire damsons are famous for their quality and flavour, but though damson trees are found in most Cheshire orchards, there are hardly any orchards which are entirely given up to this fruit. In southern and western Cheshire the damson trees grow along the roadsides in the hedgerows, and are often to be seen rising above the whitethorn hedges which separate the fields; it is therefore impossible to estimate the number of acres on which damsons are grown.

Woodlands are of various kinds. First there are plantations or woods which have been planted within the last ten years; then there are coppices, or woods which are cut periodically; and lastly the woodlands generally, which in Cheshire are mostly reserved for game and are in consequence planted with evergreens and undergrowth to give thick cover under the trees. Woodland covers 24,665 acres in Cheshire, a decrease of only about 200 acres in the last ten years, but of this area only 633 has been planted recently, and 178 acres alone is coppice. In the woods and parks there are many fine forest trees, but oaks, though plentiful, are not so large or ancient as we find them in some counties. Beeches grow to a great size on the Cheshire soil, and some of the beech avenues in the parks are exceedingly beautiful. Willows for basket-making are cultivated in marshy spots, and are cut regularly near the roots, but there are very few of the ancient pollard willows, such as we see along the river banks in the south of England.

In 1907 there were 179,827 horned cattle in Cheshire, or an average of 274 per 1000 acres, which is more than double the average for the whole of England. Most of these are dairy cattle, and the milk produced from them is largely used for the manufacture of cheese and butter, though many gallons are sent to the large towns in all parts of the country, some even being carried so far as London. The majority of the Cheshire farmers are dairy farmers, and Cheshire cheese has a well-earned reputation everywhere.

On the other hand but few sheep are reared in Cheshire. In 1907 the total was little over 99,000, and whereas the average for England per 1000 acres was 464, it was only 153 for Cheshire. In the lowlands of Cheshire the sheep are large and have heavy wool, but on the marshes and hills Scotch and other mountain breeds are grazed. Pigs, mostly "middle whites," numbered 87,391 in 1907, and horses for agricultural purposes 27,162.

14. Industries and Manufactures.

It has been already stated that Cheshire is mainly an agricultural county, but apart from its salt and chemical trade, which will be dealt with later, it has some exceedingly important industries and manufactures.

Stockport is the centre of the cotton trade of Cheshire; and indeed the Cheshire cotton trade, almost inseparable from that of Lancashire, employs great numbers of people in the eastern portion of the county. Spinning mills, weaving sheds, bleaching, dyeing, and printing works stand in all the valleys of the clear hill streams, and are thickest about Stalybridge, Dukinfield, Hyde, Romiley, and Marple. Over 40,000 men, women, and children are employed in the spinning of cotton yarns and the manufacture of cotton fabrics, which are mainly for the shipping and not the home trade. Manchester, Liverpool, and Birkenhead distribute the products of this Cheshire industry all over the world. At one time there was a

Stockport

considerable woollen industry in the Tame Valley, but cotton has replaced it, and now only about 200 hands are engaged in the woollen trade.

Before the introduction of power machinery there were numbers of hand-loom weavers in eastern Cheshire, and in all the country districts round Stockport the rattle of the loom might be heard in the cottages only thirty

years ago.   Hand-loom weaving has not been taught to
the younger generation, and as one by one the old men
and women died, the looms were broken up.   In the
neighbourhood of Altrincham and Lymm fustian-cutting
was formerly carried on in the cottages, but it has gradually vanished as the cutters were collected together into

Quarrybank Mill, Styal
(One of the early Cheshire Cotton Mills)

specially prepared mills ; indeed the cottage industries
have disappeared.

The silk trade of Macclesfield started in 1756 and,
though suffering from many ups and downs, is now of
great importance ; in fact, Macclesfield is still at the
head of the British silk industry.   It is also carried
on to a lesser degree at Sandbach and Congleton, at

which latter town a silk-mill was opened in 1755. The ancient trade of Macclesfield and Congleton was in buttons, twist, bobbins, and tags and laces, or points as they were called. Nearly 9000 people are now engaged in the silk industry.

The hat trade of Stockport, Hyde, and Bredbury employs 7320 hands, and the kind of hat produced varies according to the fashion of the day, silk having replaced beaver, and felt silk. Straw hats are also made, though not in such large quantities.

The leather industry still employs a large number of Cheshire workers. Tanneries are not so plentiful as they were a few years ago, but there are some at Birkenhead, Nantwich, and Stockton Heath, and a few in country places. The boot and shoe trade of Nantwich and Sandbach is an ancient one, and is followed by some 2400 persons. Near the former town it is interesting to see leather-paved footpaths. The rejected scraps of leather from the boot warehouses are used by the council for this purpose and are certainly cleaner than cinders.

Upwards of 30,000 people are employed in various branches of the iron trade, but it is difficult to separate this complicated business into classes. There are important engineering and iron works at Chester, Birkenhead, the neighbourhood of Warrington, and Northwich. Within the last few years an industrial colony, where engineering works of various kinds employ many hands, has arisen at Broadheath near Altrincham. There are cable works at Helsby, and road-motor works at Sandbach, and a large

Congleton

number of agricultural implements are turned out in the county.

Ship-building, which now is a branch of the iron trade, finds employment for large numbers of hands at Birkenhead and along the Mersey shore, and the building of flats and smaller craft is carried on at Northwich and Runcorn. From Laird's yard at Birkenhead sailed, in 1829, one of the first iron-built steamers, and in 1862 the blockade-runner "Alabama," which cost the country so much owing to her warfare with the Northern States of America, was built in the same yard. Ship-building was carried on at Ellesmere Port before it was started in Birkenhead.

Amongst minor industries we may refer briefly to jam-making at Stockport, margarine works at Godley, soap-making at Port Sunlight and Winsford, and brewing at Chester, Stockton Heath, and elsewhere.

15. The Salt and Chemical Trade.

The salt trade of Cheshire has great antiquity. Though there are no actual records, it is probable that even the British inhabitants of the county obtained salt from the brine springs at Nantwich and Northwich, both of which are said to have had Keltic names. There is ample evidence that the Romans encouraged the trade at the springs, and at Nantwich there were eight salt-houses when *Domesday Book* was written. The Welsh knew Northwich, the "black salt town," and Nantwich, the

A Fitting-out Basin, Tranmere, Birkenhead

"white salt town" well, for they carried on a trade with them in very early days. Salt obtained from the Cheshire salt-springs has always been a typical Cheshire trade.

Brine springs are found all round Northwich, Wharton and Winsford, Middlewich, Nantwich, and on the border at Wigland near Malpas. The salt-pans at the last place ceased to be used about 1850, though salt had been worked at "Dirtwich," as it was called, since the days of Edward III; and at Nantwich the trade has practically vanished, though the brine baths are now used for curative purposes. In the neighbourhood of Winsford and Northwich, however, the trade in salt obtained by evaporation has steadily increased, and now brine is pumped over a wide area.

When in 1670 rock-salt was discovered at Marbury near Northwich the salt trade received new impetus, and salt mining largely took the place of salt evaporation. The rock-salt is boiled and cleared. It must not, however, be thought that brine pumping has in any way gone out; brine is now carried in pipes for miles from the springs to the salt-works. Between seven and eight thousand hands are employed in the salt and alkali trades, and many of the firms engaged have branches for the manufacture of soap.

More than fifty years ago 600,000 tons of white salt and 120,000 tons of rock-salt were carried from Northwich and Winsford down the Weaver. Twenty years later, in the year 1871, over 1,000,000 tons were shipped from Cheshire. In 1904 the quantities were 53,780 tons of white, and 1,187,442 tons of rock-salt obtained, but the shipping trade in pure salt has fallen off within the last few years.

The alkali and other chemical trades of North-wich, Middlewich, Sandbach, Lostock Gralam, Runcorn, Weston, Stockport, and Frodsham Bridge are a late

Winsford Bottom Flash

development of the working of the saliferous earths of central Cheshire, which has steadily risen in importance and almost swamped the trade in pure salt.

Before leaving the subject it is necessary to speak of one great drawback of the salt trade, namely the conse-quent subsidence of land over the brine springs. It is the custom, when the rock-salt is mined, to leave supporting pillars of salt; when sufficient rock-salt has been excavated, the levels are flooded and the dissolved salt pumped out. As the supporting water is removed the land gradually, and not always gradually, sinks, with the consequence that buildings are thrown out of the perpendicular, "flashes" are formed, and unfortunately serious damage is occasionally done to property. In Northwich,

A Subsidence in Northwich

where the worst subsidences have taken place, the difficulty is now done away with by the skill of the engineer and architect. The houses are built on strong wooden bases, and enclosed in timber frameworks. When subsidence occurs the whole building is raised by jacks, and brick courses are introduced beneath it, the roadways, which suffer considerably, being also raised. A few years ago there was one house on the Warrington Road which stood some feet above the roadway, its front door being approached by steps; a few years later it was possible to look in at the bedroom windows from the footpath; the house has now been taken down. Some large and important buildings, when their foundations became unsafe, were bodily moved to a new site, a custom employed occasionally with timber buildings in America but seldom known in England. Many thousands of tons of waste products are tipped into the big flash known as Witton Brook or Northwich Mere, but the subsidence still grows.

Practically the salt and chemical trades are confined to the Weaver valley, and in places the smoking chimneys, steaming brine-pans, and wooden erections above the pumping stations give this portion of the county an appearance which may be likened to the Black Country of Staffordshire or South Wales, but a mile or two away from the river on either hand, even close to Northwich itself, is some of the most beautiful country in Cheshire. Considering the value of Cheshire salt-bearing earths and rocks, we can afford to spare a little of its beautiful scenery.

16. Minerals—Past and Present Mining Industries.

It will have been noticed that to obtain the rock-salt from the Cheshire marls mining is necessary, but in addition to this there is a considerable colliery industry

C. C.   7

in Cheshire. In the chapter on Geology it was stated that the coal was chiefly obtained in the east of the county, and that there were mines at Neston in Wirral. We need not say more about this industry except to state that there is an output of about 400,000 tons of coal annually, and that in 1904 its value was nearly £150,000. Cheshire coal is not, as a rule, carried to any great distance, but is used in the various factories in the north-east part of the county. Over 3200 persons are engaged in the coal trade.

The returns from the quarries in 1904 were 116,433 tons of sandstone for building purposes, while 14,716 tons of fire-clay and 68,386 tons of brick-clay were raised. Over 24,000 tons of gravel and sand were also dug up for building purposes.

A very ancient and interesting mining industry existed at Alderley Edge in a sandstone bluff that rises above the plain. The discovery of certain primitive mining-tools led to the careful investigation of the Edge and the mines, which resulted in conclusive evidence that the mines were worked in Roman days, and that as the Edge was probably peopled earlier it is possible that mining of some sort was carried on before the Romans discovered the place. Lead and copper were obtained from the sandstone at Alderley and Mottram St Andrew by the Romans, and by the British working for them with rude stone hammers. The ore was smelted in open-air hearths close to the mines.

Whether mining was continued long after the Romans had left we do not know, but it is suggested, from certain place-names, that it existed in the sixteenth or seventeenth century. At the beginning of the eighteenth century the mines were reopened, but the first prospector was ruined. About 1755 Charles Roe, who founded the Macclesfield silk trade, set to work and started a company, employing about 40 or 50 miners, making a clear profit of about

Alderley Copper Mines

£50 per week. In 1770, however, the company moved to the more profitable Parys Mountain in Anglesey, taking all its miners with them.

Lead and copper ore were obtained by various individuals and small companies during the next hundred years, and cobalt was discovered and worked for a time with success. The Alderley Copper Mines in 1867 produced 15,152 tons of copper ore, containing about 2 per cent. of metal, valued at about £75 per ton. In 1865 the Magnesium Metal Company worked the Mottram mines for lead, copper, and cobalt. In 1879 the Alderley mines were closed and the industry was over.

Lead and copper were worked for some time at the southern end of the Peckforton Hills, but the industry was never of importance; the ore obtained did not justify continuance of the works, which are now ruined and deserted. Attempts to find minerals in the Wirral hills and elsewhere in the sandstones have not met with success.

17. Fisheries and Fishing-Stations.

The fisheries round the coast of England are of great importance, for they employ large numbers of people and provide food for the inland districts as well as the coast towns. The actual coast-line of Cheshire being short, we should expect to find that in the county the fishing industry is not considerable. Only about 300 fishermen sail from Cheshire ports, but this number does not include the Handbridge fishermen of Chester.

Fishing is carried on in salt and estuarine waters in various ways, by trawling, line-fishing, and stake-netting, and also by "tees" and trammels. The chief centres are Hoylake, Parkgate and Heswall, Tranmere, Rock Ferry, New Brighton, Egremont, Ince in the estuary of the Mersey, and Handbridge by Chester, on the Dee.

The Hoylake fleet of trawlers is the most important, but there are fewer boats now than there were when the anchorage in the Lake was more commodious. It

A Hoylake Trawler

consists of 39 "first-class" trawlers and 12 inshore fishing boats. The first-class boats are yawl-rigged smacks of about 40 tons register, and some 60 feet in length ; they carry beam trawls of about 50 feet spread, and have steam capstans for hauling in the trawl and hoisting the sails. They fish in deep water off the coasts of North Wales, the Isle of Man, and Lancashire, and chiefly get plaice and soles. Second-class boats sail from the other ports. There are 36 trawlers and open boats at Parkgate and Heswall, and over 30 at the Mersey ports. The smaller trawlers are half-decked, about 10 tons register, and carry a trawl of about 25 feet spread; they fish for "flukes" and shrimps in shallow water. The open boats are mostly used for line-fishing and in the mussel fishery. In 1908 over 5500 tons of fish, valued at £5600, were landed at Birkenhead and Hoylake alone. This, however, does not represent anything like the whole of the fishery, for the Hoylake trawlers land their catches at Liverpool, Douglas, and Bangor as well.

The sea-fisheries are regulated by the Lancashire and Western Sea Fisheries Committee and the Dee Sea Fisheries Committee, which are supported by the various County Councils. The freshwater fisheries in the Dee are managed by the Dee Board of Conservators. The committees regulate the areas over which certain kinds of nets may be used, the meshes of nets for certain fish, the size of fish which may be taken, and various other matters connected with the capture of fish. The Lancashire and Western Committee has a large area, including the coasts of Cumberland, Lancashire, and most of Wales, and the fishing-boats from various places trawl and line-fish within its area.

The shallow waters off the coast of Cheshire are what are termed nurseries for many species; that is to say they abound with the young of plaice and various flat-fish, and often with the young of sprats, herrings, and other food-fish. It is therefore very necessary that there should be strict regulations to prevent the unnecessary destruction of these little fish while they are still too young to have a market value. Shrimp trawling, largely carried on in the Dee, is responsible for the death of countless thousands of them, for even if they are thrown overboard immediately they are brought up in the net, they are often crushed and killed by the weight of the other fish caught with them.

Trawl-nets are mostly used on the Cheshire banks and channels, the trawl being dragged along the bottom. Stake-nets, not so much used now as in former years, are walls of net set up on stakes at low water and examined immediately the tide has gone down ; these nets are not usually set parallel to the shore, but are arranged so as to suit the set of the tide over the banks. In the Dee drift-nets are used from boats, and occasionally trammels are employed. The trammel is a three-fold net, the outer walls are large-meshed, and the inner net of smaller mesh ; it is weighted at the lower edge and buoyed at the upper, so that it stands vertically in the water. Fish ascending or descending the river pass through the first large mesh and push the narrow-meshed net through the meshes of the third, enclosing themselves thus in a pocket until taken out. Line-fishing is also carried on, but not to any great extent.

The banks of Liverpool Bay and the estuaries are great places for shrimps, prawns, and cockles. The two former are captured either with trawls from a sailing-boat, or with wide nets which are pushed through the shallow water by the fishermen. Men, women, and children may often be seen pushing these big nets along the edge of the tide a mile or more away from the shore.

"Scratching" or raking for cockles is an old industry at Hoylake and Parkgate, and still finds employment for large numbers of fisher-folk of both sexes. All shrimps, prawns, and cockles have to be riddled, and the small undersized crustaceans and molluscs fall through and are not destroyed.

As over 120 different kinds of fish are found off the Cheshire coast, it is quite impossible to name them all here. The most important food-fishes are the plaice, sole, flounder, dab, halibut, brill, turbot, whiting, haddock, cod, and sprat. Two small fishes, of no value for food, are found abundantly in the shrimp-trawls; these are the dragonet, one of the most beautifully coloured of all our fishes, and the lesser weever. This last fish is very troublesome, for its spines are poisonous, and a careless shrimper may be seriously stung. Shrimpers avoid sorting their catches at night on account of these fish, but in the day-time they are very expert at picking them out and throwing them overboard before the fish have time to strike them. Herrings and mackerel occasionally come inshore in great numbers, and the herring-boats have on several occasions followed the fish into the Mersey.

The Dee is a salmon river of importance, and has

fishing regulations distinct from those of the Committee. The fish are captured by strike-nets floated from a boat, by coracle nets, and by rod and line. The best fish are captured in the straight cut below Chester by the Chester fishermen who live at Handbridge. Sturgeon are not infrequently taken in the Dee, and some curious

Salmon Netting on the Dee, below the Old Dee Bridge

formalities have to be gone through by the fishermen before they may sell the Royal fish. Dogfish are sometimes very troublesome, taking fish off the hooks on long-lines, which are baited and left down for some hours, and occasionally the fishermen's nets are broken by the larger species of dog-fish, often called sharks,

which get entangled in them. Skates are not uncommon, and the curious devil-fish or angler is at times
stranded on the banks.

The Mersey was once a salmon river, but the pollution
of its upper waters and the chemical refuse in the Weaver
have driven away the fish ; nevertheless so lately as 1908
a salmon was killed at Warrington.

18. Shipping and Trade. Chief Ports.

Birkenhead is the largest port in Cheshire, but it can
hardly be separated from Liverpool. Its great series of
docks are in the Great Float, formerly Wallasey Pool, an
arm of the Mersey. Round this float, which empounds
the tide so as to be always full, are $9\frac{1}{2}$ miles of wharfage,
an area of 506 acres of docks, and 165 acres of water
deep enough to float any ocean steamer. Roughly the
Birkenhead docks cover one quarter of the total area
under the Mersey Dock and Harbour Board. In addition
to these there are private docks and wharves belonging to
the railway companies.

The Birkenhead Stage is a long floating raft, 800 feet
by 80 feet wide, connected by three bridges with the
land, one of which is 678 feet long. The Wallasey stage
is only 200 feet shorter. The ferry between Woodside
and Liverpool carried 9,294,201 people in 1904, and the
other Wallasey ferries, 17,209,537. These big figures
mean that $26\frac{1}{2}$ million passengers were carried across the
river, though of course some of them went every day in

the year.   We must remember that this by no means represents all the traffic between Liverpool and Birkenhead, for the towns are also connected by an electric railway which is tunnelled under the river.

A very large proportion of the trade in the Birkenhead docks is in American cattle, which are brought in alive and often taken straight into the abbatoirs to be killed.

Runcorn Docks

In 1904, 554,620 head of cattle and sheep were landed at Birkenhead.   Regular lines of steamers run from Birkenhead to India, China, and South Africa.

Chester, once a far more important port than Liverpool, is now only used by small coasting vessels and flats.   All the other Cheshire ports are either on the Ship Canal or the Weaver.

Ellesmere Port, at the junction of the Shropshire Union Canal with the Manchester Ship Canal, has wharfage accommodation for many ships and flats without impeding the traffic on the Canal. It is mainly engaged in the Staffordshire iron trade. Northwich and Winsford are ports in the chemical and salt trade, and the boats which bring the goods down the Weaver mostly tranship their cargoes to ocean-going vessels at Weston Point, the Port of Runcorn. There is here accommodation in docks and wharf for a large number of vessels. Partington coal basin covers $6\frac{1}{2}$ acres of water space, and has a quay half a mile in length ; it is connected with the railways bringing coal from the Lancashire coal-fields. This is the last of the Cheshire ports, but along the Ship Canal which runs through Cheshire land may be seen the great ocean steamers going to and from the now growing port of Manchester.

19. History of Cheshire.

The early history of Chester and the county of Cheshire is founded more on tradition than fact, and we really do not know the exact date when Cheshire came under the sway of the Romans. It is certain that no Roman army succeeded in penetrating to the British stronghold of Caerlleon or Deva in the invasions before the days of Christ, but it is possible that Cheshire was visited by the army of Claudius a few years after the third invasion of England in 43 A.D., probably when attacking the Britons in North Wales, and when on an expedition against the Brigantes, whose territory lay north of the Mersey.

Certainly Cheshire was occupied and probably some of its roads made by 80 A.D., but the actual origin of the Roman camps at Chester, Northwich, Middlewich, and other places is not really known. During the Roman sway, however, which terminated about 410 A.D., all the land now known as Cheshire was completely under the Roman rule, and we shall have occasion to speak of the Roman antiquities of the county in the next chapter. A Roman legion—the XX, known as Valens Victrix (mighty and victorious)—which consisted of some 6000 infantry and 400 cavalry, was stationed at Chester, it is said, for two centuries, if not more.

No sooner had the Roman legion left Chester unprotected than the Welsh attacked the civilised Romano-British city and once more gained possession. For nearly two hundred years Cheshire was at the mercy of raiding Picts from Scotland, and piratical Scots from Ireland, but during that time Wirral is said to have received a missionary visit from Ireland, while one of the first Christian settlements was founded in Cheshire. Tradition has it that St Patrick sailed from the Mersey for Ireland in 432 A.D., but tradition also says that he landed from Ireland in Bromborough Pool, where he blessed a well. Whichever be true, it is, however, certain that there were very early Christian missions in Wirral.

Aethelfrith, king of Northumbria, captured Chester in 607, and revenged himself by slaughtering the monks of

Tombstone, XX Legion, found in Chester 1891

(*Coll. Grosvenor Museum*)

Bangor, who had come to pray for pity, to the number of about 1200, according to Bede, though the Saxon chronicle says only 200. Shortly after that we find the Britons again in Chester, where they were once more defeated when, about 813, Egbert king of Wessex conquered Mercia.

In 894 the Danes invaded and captured Chester, but were defeated by King Alfred's troops, and the "Saxon Amazon" Aethelfleda, Alfred's daughter, is said to have built a town on Eddisbury Hill in 914, and castles at Warburton and at Runcorn in the following year; these were defences from the northern invaders, the two last castles being on the Mersey.

King Edward the Elder built the fortified town of Thelwall on the Mersey in 923, and "sat there" while other Mercians were crossing into Lancashire to attack and hold Manchester in Northumbria.

The Danes, in spite of this defeat, do not seem to have been finally driven out of Wirral, and they had settlements at several places in that part of Cheshire. We may notice that in Wirral there are a number of villages with the Danish termination "by"—for instance West Kirby, Irby, Frankby, and Raby. There is a tradition that King Edmund drove the Danes out in 947, and when Edgar visited Cheshire that six kings did homage to him and rowed him in state upon the Dee.

Peace did not last long, for shortly after this we find Aethelred the Unready fighting with the Danes, and in 1014 Canute the Dane king of England. Chester and all the country round suffered from the ravages of the contending forces of Canute and Edmund Aetheling, but whether or not Canute himself ever passed through Cheshire and crossed the marsh at Knutsford or Canute's ford, we shall probably never know.

When William the Conqueror had defeated Harold, last of the Saxon kings, at Hastings, he divided the country into earldoms, and the Saxon Earl Edwin submitted to the Norman rule and for a time remained in possession of this portion of Mercia. In 1069, helped by the Danes, Edwin and Morcar rebelled, and William marched north carrying destruction; he laid the land waste with fire and sword, finishing in 1070 by handing over captured Chester to his fierce soldiers to pillage. In *Domesday Book* we read that after this invasion a great part of Cheshire was "waste"; it is evident that the Cheshire men did not submit tamely, and suffered for their stubborn resistance. Giraldus Cambrensis states that Harold was not slain at Hastings, but retired to Chester wounded; the spot is shown where he is said to have died as a hermit.

The Earldom was at first handed over to Gherbod, one of William's leaders, but he did not remain long in the hostile land, and Hugh Lupus, William's nephew, was really the first Norman Earl of Chester. He was given a free hand to rule as he liked, and his sign of office was a sword. Hugh Lupus and his eight Barons, if all the stories about him are correct, indeed ruled Cheshire by the sword.

In the twelfth and thirteenth centuries Chester and western Cheshire were in the uncomfortable position of

Chester Walls and the Dee

being the scene of constant conflict between the English and the rebellious Welsh. Armies, frequently led in person by the kings, were repeatedly in the town or encamped against it. Henry II, John, and Henry III led armies against the Welsh, but got no further than Chester. About 1245 Henry III caused all the brine-pits at Nant-wich to be destroyed, and a few years later laid waste all the Cheshire border-land to take temptation from the invading Welsh. In 1276 Chester was the base of the successful expeditions of Edward I, and in the following year he laid the first stone of the Abbey of Vale Royal with pomp and ceremony. Wales finally submitted in 1282, when Edward was again at Chester, and in 1300 we find the Welsh chieftains at Chester doing homage to the Prince of Wales.

The Black Prince was in Chester in 1353, and Richard II when on his way to Ireland in 1394. During the Wars of the Roses Cheshire men were divided, and at the battle of Blore Heath the knighthood of the county was almost exterminated in civil war. In 1513 Cheshire men figured again prominently at Flodden, for indeed during all the mediaeval wars they were constantly to the front.

In 1617 James I made a royal visit to Cheshire, hunting in Delamere Forest, and entering Chester in state. Twenty-five years later some of Cheshire's famous gentry had risen against the Stuart rule, and Sir William Brereton caused a drum to be beaten for the Parliament in Nantwich. Then for three years the county was once more divided ; loyal Chester was the hope of Charles,

Nantwich the stronghold of revolution. Nantwich, Beeston Castle, and Chester were scenes of conflict, sometimes sieges, sometimes continuous guerrilla warfare, until Brereton moved his camp to Dodleston and surrounded the city of Chester. The battle on Rowton Heath, within sight of the king from his post of safety on Chester walls, shattered his hopes in Cheshire, and very shortly the city fell.

Cromwell, with 10,000 men, held the bridge at Warrington against the Duke of Hamilton's advance from Scotland in 1648, and turned the advance into retreat. After the battle of Worcester in 1651, retreating Scottish troops passed in disorder through Cheshire, and over 100 were taken prisoners by the men of Sandbach, when, weary and disorganised, they passed through the town on market-day. Sir George Booth, a native of Cheshire, raised 3000 men in 1659 in rebellion against the Parliament, but Lambert defeated him at Winnington near Northwich. This was practically the last battle on Cheshire ground.

In the rebellion of 1745 Charles Edward passed through Cheshire, and there are many stories and not a few relics of the visit of the wild Highland troops of the Young Pretender.

When Napoleon was threatening invasion, in 1804, the Duke of Gloucester held a grand review of regulars and volunteer troops on the then wide plain of Sale Moor, an event well described by Mrs Linnaeus Banks in *The Manchester Man*. In 1832 the Princess Victoria opened the far-famed Grosvenor Bridge at Chester.

In later years Chester was nearly the scene of a great disaster. A Fenian plot, in 1867, arranged the seizure of Chester Castle, and an organised dislocation of the railway and telegraphic communication until the arms and ammunition from the depôt had been removed to Ireland. Between 1400 and 1500 Irishmen gathered without the city, but news of the plot had leaked out, and the military and special constables were prepared. The hostile crowd dispersed and the danger was averted without conflict.

Cheshire, we see from this sketch, has frequently figured in English history, but it is impossible to relate in a short chapter how frequently the men of Cheshire have been to the front in the deeds which made our land and Empire what it is.

William the Conqueror himself visited Chester in 1069, and it was in the following year that his nephew Hugh, best known as Lupus or the wolf, was made Count Palatine of Chester. Thus Cheshire was the first County Palatine or Royal county, a dignity it retains to the present day, for our Prince of Wales is also Earl of Chester.

20. Antiquities—Prehistoric, Roman, Saxon.

In Cheshire, as in other parts of England, we cannot depend upon written records of the earliest people who inhabited the land, but have to search for other evidences of their habits and history. In many places there are ancient graves which have been opened, or buried villages which have been explored, and relics of different kinds, both bones of the early inhabitants and the implements and weapons they used, have been found and studied by archaeologists.

These students of early history have roughly classified the men who lived in prehistoric times, or in the ages before the Romans came, into various periods or races, but in each we find traces of men of different origin. The earliest of these ages is called the Stone Age, when the very roughest and rudest implements of stone were in use. The next is the Bronze Age, when men were beginning to make use of metals, obtaining the ores of copper and tin from the rocks and smelting them into bronze. The last is the Iron Age, when the methods of working iron and the great advantages it possessed over the softer metals had been discovered, and sharper and more useful weapons and implements were made. Men of the Iron Age were in Britain when the Romans arrived.

Only very rough guesses can be made at the length of time that the different races, or men of various ages, inhabited Britain, but we know that there were two very distinct races in the Stone Age—the Palaeolithic or older, and the Neolithic or later race. The shape of the skulls and the stature of these two were quite distinct, but what is more noticeable is that Palaeolithic man used much rougher stone implements than the other, and that these early Stone men did not seem to have any knowledge of pottery, while Neolithic man often ground and polished his weapons.

In Wales, which being a mountainous country was always the refuge of races which were being driven back by invaders, we find that both races were represented, and those who have made a study of the question can show us that the older race, whom they call Iberians, for they are supposed to have migrated from Southern Europe, survived for a long time in these mountain fastnesses, even after the later Goidels or Kelts, who had much rounder skulls, had come amongst them. The influence of the two races is visible in the small dark, and the red-headed round-skulled races we find in Wales to-day.

Both these early Stone men raised mounds or barrows over the graves of their dead, but the early men made what are called long barrows and the later men round barrows. In Cheshire, which we must remember was always the site of conflict between the invaders and defenders of the land, being just on the edge of the hill-country to which the defeated were driven, there are hardly any traces of the very early race, but many relics of the later Stone men. When the men of the Bronze Age had learnt how to make metal weapons they had great advantage over those who only used rougher weapons of stone and wood, and they came, probably over the hills of Derbyshire, and drove the Goidels westward. The race which left most traces in Cheshire were the Brythons, of which the Cornavii were a tribe inhabiting the level land between the Mersey and Severn at the time when the Romans brought civilisation into Cheshire.

It is very hard even for men who have made a life-long study of the relics left by these early men to be quite

Palaeolithic Flint Implement
(From Kent's Cavern, Torquay)

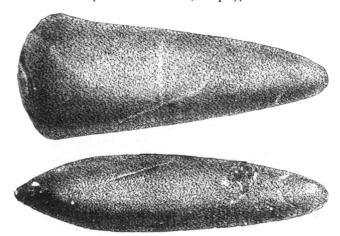

Neolithic Celt of Greenstone
(From Bridlington, Yorks.)

sure to which age some of the burial mounds and the
instruments found in them really belong.   We must
always remember that it takes a very long time for men
to give up using primitive implements when it is difficult
to get improved ones.   There have been polished stone
axes and celts dug up in Cheshire, and very much rougher
tools of a later date found also, and at first these rough
tools were thought to be relics of the earliest days.

We may mention here a few of our more important prehistoric remains.   The most perfect stone circle in Cheshire
is perhaps the one which stands close to the much more
modern " Cross " at Clulow Cross, on the hills above
Macclesfield.   These stone circles were evidently used
for early religious rites, and in some cases were also
burying places of chiefs or men of authority.   Not far
away, on the top of the hill behind Bosley Cloud, and
right on the Staffordshire border, there is a very famous
circle, the Bride Stones, which is not so perfect a circle but
is in better preservation than the one at Clulow.   There
are others which may be traced in the neighbourhood of
Alderley.

Although circles and Neolithic standing stones, or
menhirs, as they are called, are rare in Cheshire, there
are large numbers of the burial mounds or tumuli.   These
are mounds varying in size, usually placed high up on the
hills.   They are most numerous in two groups, one in the
eastern portion of the plain and the other amongst the
Delamere hills.   There are three on the hills at Rainow,
and five in the neighbourhood of Capesthorne, though
many miles apart.   In Delamere we find them as far

apart as Castle Cob in the north and the Seven Lows near Oulton; and there are others in Dunham Park, at Warburton on the Bollin, at Eccleston on the Dee, and near Tarporley. Some have been opened and yielded fragments of funeral urns containing charred human remains, flint spear- and arrow-heads, and other Stone Age relics. At Lymm some primitive axe-hammers of red-deer horn have been found, and elsewhere in the

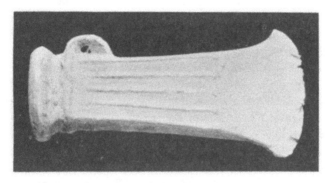

Bronze Axe from the Manchester Ship Canal
(Coll. Grosvenor Museum)

county celts, hammers and flint scrapers, borers, and other implements have been discovered.

At Alderley, Delamere, Stockport and elsewhere bronze celts, fragments of weapons, and pottery of the Bronze Age have come to light, but the most numerous relics of primitive man are referable rather to the Iron Age men who inhabited the country prior to the Roman conquest. Some of these remains come from the submerged forest at Dove Point, but the localities whence they have been obtained are scattered all over the county. Probably most of the bones of domestic animals found in the Forest Bed and elsewhere are of the flocks and herds of the pastoral Cornavii.

The terminations of place-names are frequently in-dications of the former inhabitants, and the word "Low" mentioned above, which is derived from "hlaw"—a burial mound—is to be met with in many places where no tumulus now exists, for instance, Bucklow, Wilmslow, and Twemlow.   Lindow, the name of the moss near Wilmslow, is also Keltic; it is *Llyn-dw*, or the black lake, which formerly lay above the present moss.

There are a few camps on the Cheshire hills which may still be traced, though in most of them the ditches have been filled up and the earthworks levelled.   It is, however, very difficult to fix the dates of these old earth-works, because they have been used at later dates than early British days.   When the Romans were marching through the country they were obliged to attack these hill encampments, and since the only way to protect the roads they were making through the forests was to occupy the camps, they turned the British out, repaired, and re-garrisoned them.   The British or the Picts and Scots were in them when the English came, and later the Saxon invaders repaired the captured strongholds and occupied them.   It is therefore not surprising that we find British, Roman, and Saxon remains near these camps.

Maiden Castle, as it is called, on Bickerton Hill, the

southern hill of the Peckforton range, is at once one of the best preserved and the most carefully selected of the hill forts. The ditch and rampart guard it on two sides, the steep escarpment and slope on the other. Kelsborrow Castle guards the pass over the Delamere hills, an old British road, but very little now remains except the position and the name. Eddisbury Hill is crowned by a well-marked camp, the age of which is uncertain ; it was repaired and not built in Saxon times. Bucton Castle, on a steep hill overlooking the Tame near Stalybridge, was a fortress in the east of the county, and there is an exceedingly well-defended promontory at Oakmere, though the present mere does not cover so much ground as it did then.

When the Romans occupied Cheshire they had camps at Chester (Deva), Kinderton (Condate), and Wilderspool (Veratinum), though the last was an outpost only, protecting their manufacturing town. Beside this they had several less important stations guarding the roads, but we shall speak of these Roman roads in a later chapter. In the neighbourhood of these camps and outposts very many interesting Roman relics have been found. Chester was a most important Roman town, and the remains of the dwellings, municipal buildings, baths, forts, and manufactories are not only numerous, but show how wonderfully the conquerors civilised the place and introduced their habits and arts. It is said that Chester walls were built by a British king, which is quite possible, but the oldest walls which can be found to-day are those which were undoubtedly built by Romans, and we even find that the

Remains of Roman Hypocaust and Chester Walls

Roman water-supply of Chester was carried in lead pipes. It is thought that the walls of Chester were built about 84 A.D., by Agricola, for there is no foundation for the tradition that Julius Caesar penetrated to Chester.

The Romans very soon realised that the brine springs were worth defending and working, and Northwich, Nantwich, and Middlewich were fortified and occupied. In these places we find many Roman remains, but

Roman Wall, St John Street, Chester

perhaps the most interesting are at Wilderspool on the Cheshire side of the river at Warrington. When, a few years ago, this town, which must have been a busy hive of industry in Roman days, was discovered and excavated, traces of various manufactures were found, amongst them potter's kilns (in which they made imitations of Samian ware), iron and glass furnaces, the workshops of enamellers and workers in jewels. The station was on the road from Wigan to Condate or Kinderton, and Wigan coal, as well as some from Wales, was found. The Roman relics in the Chester and Warrington museums are well worth studying.

Stone Hammer, Alderley Edge

(*Coll. Grosvenor Museum*)

There are some Roman antiquities which are very interesting, for they were a puzzle to antiquarians for a long time. We have said that the men of the Stone Age made stone hammers, and these tools were also used by men of the Bronze and Iron Age to break up the sandstones from which they got their copper and lead. At

Alderley Edge large numbers of very rough stone hammers have been found; they are mostly boulders picked up from the glacial deposits on the plain, and grooves were rubbed round them so that they could be attached to wood shafts by thongs or withies. These Alderley stone hammers were used not by the men of the early ages, but by the British slaves mining for copper and lead for their Roman masters.

The Romans did not obtain all their lead from Alderley, but had a regular trade in lead with the Britons in Wales. Some exceedingly interesting lead pigs, about 2 feet long, and 4½ cwt. in weight, were dredged from the Dee at Chester. They had probably fallen into the river during transhipment. So advanced was this trade that we actually find that the British tribe from whom these pigs were obtained advertised their wares in the way we do now by stamping them with the name of the manufacturer. A very perfect example of this kind is preserved in the Grosvenor Museum at Chester.

The most interesting of all the Saxon relics in Cheshire are the early Christian crosses in Sandbach market-place. These had been thrown down, built into walls and pavements, carried in fragments to various places in the neighbourhood, and apparently lost and destroyed. Ormerod, the Cheshire historian, collected all the scattered fragments which could be found, and had them put together and re-erected in their original situation. The actual history of the crosses is not known, but it is suggested that they were put up to commemorate the marriage of Peada, son of Penda of Mercia, to the daughter of Oswi of Northumbria. Another interesting relic is a runic stone which was found at Upton in

The Crosses at Sandbach

Wirral, and many Saxon stones have been found in different localities.

Cast of Runic Stone found at Upton, Wirral

(*Coll. Grosvenor Museum*)

21.   Architecture. (a) Ecclesiastical— Cathedral, Churches, Abbeys and Monastic Houses.

The Architecture of our land may be conveniently divided into three classes, first that connected with the Church or ecclesiastical buildings; secondly the military forts and castles ; and lastly the domestic dwellings, houses and cottages.

If we go into any part of England and look at the buildings and houses we shall see that they are made of stone, wood, or brick, and the kind of stone used depends to a great extent upon the material which is most easy to obtain, though of course in modern times building materials for large houses or public buildings are often brought from a great distance when there is no very lasting material to be obtained locally.   In Cheshire we have sandstone, which is good for building purposes if the right kind is quarried; we shall see that this has not always been the case.   Nowadays the clay of the lowlands, baked into bricks, is largely used throughout the county, but this too has not always been the case, though clay was one of the earliest materials used for dwelling houses.

A preliminary word on the various styles of English architecture is necessary before we consider the churches and other important buildings of our county.

Pre-Norman or—as it is usually, though with no great certainty termed—Saxon building in England was the work of early craftsmen with an imperfect knowledge of stone construction, who commonly used rough rubble walls, no buttresses, small semi-circular or triangular arches, and square towers with what is termed "long-and-short work" at the quoins or corners.   It survives almost solely in portions of small churches.

Pulpit in Refectory, Chester

The Norman Conquest started a widespread building of massive churches and castles in the continental style called Romanesque, which in England has got the name of "Norman."   They had walls of great thickness, semi-circular vaults, round-headed doors and windows, and massive square towers.

From 1150 to 1200 the building became lighter, the arches pointed, and there was perfected the science of vaulting, by which the weight is brought upon piers and buttresses. This method of building, the "Gothic," originated from the endeavour to cover the widest and loftiest areas with the greatest economy of stone. The first English Gothic, called "Early English," from about 1180 to 1250, is characterised by slender piers (commonly of marble), lofty pointed vaults, and long, narrow, lancet-headed windows. After 1250 the windows became broader, divided up, and ornamented by patterns of tracery, while in the vault the ribs were multiplied. The greatest elegance of English Gothic was reached from 1260 to 1290, at which date English sculpture was at its highest, and art in painting, coloured glass making, and general craftsmanship at its zenith.

After 1300 the structure of stone buildings began to be overlaid with ornament, the window tracery and vault ribs were of intricate patterns, the pinnacles and spires loaded with crocket and ornament. This later style is known as "Decorated," and came to an end with the Black Death, which stopped all building for a time.

With the changed conditions of life the type of building changed. With curious uniformity and quickness the style called "Perpendicular"—which is unknown abroad—developed after 1360 in all parts of England and lasted with scarcely any change up to 1520. As its name implies, it is characterised by the perpendicular arrangement of the tracery and panels on walls and in windows, and it is also distinguished by the flattened arches and the square arrangement of the mouldings over them, by the elaborate vault-traceries (especially fan-vaulting), and by the use of flat roofs and towers without spires.

Chester Cathedral: Interior

The mediaeval styles in England ended with the dissolution of the monasteries (1530-1540), for the Reformation checked the building of churches. There succeeded the building of manor-houses, in which the style called "Tudor" arose—distinguished by flat-headed windows, level ceilings, and panelled rooms. The ornaments of classic style were introduced under the influences of Renaissance sculpture and distinguish the "Jacobean" style, so called after James I. About this time the professional architect arose. Hitherto, building had been entirely in the hands of the builder and the craftsman.

Chester Cathedral, which will be spoken of more fully later, was chiefly built in the fifteenth and sixteenth centuries, and is in the Perpendicular style. It was constructed of Cheshire sandstone without regard to the fact that certain strata of this stone will not withstand the influence of weather, and much of the external ornamentation suffered, and had to be restored.

Most of the older Cheshire churches were built of sandstone, and possess massive square towers erected in the sixteenth century. It would be impossible to mention all these, but very good examples may be seen at Woodchurch, where the tower has great buttresses, at Malpas, Acton, and Shotwick. Astbury is an ancient church, and contains work of all times from Norman days. It has an Early English chancel, and battlements and clerestory built at the end of the fifteenth century, whilst its ceiling and furniture are Jacobean. Its tower is surmounted by a steeple. Wybunbury Church has also a steeple above its ancient tower, but it is not a hundred

Chester Cathedral

years old; the old spire was so crooked that there was a Cheshire proverb, "as crooked as Wembury steeple," and it ultimately had to be rebuilt.

In some of the older churches there is no tower or spire, but merely a little bell-cote, often of wood. This brings us to another class of Cheshire church very characteristic of the county which can hardly be spoken

Malpas Church

of as of any particular style, unless it is early Saxon, though as a matter of fact these churches were mostly built in the fourteenth century. Most of the Saxon buildings, both ecclesiastical and domestic, were made of timber and mud. The framework was of timber—in Cheshire usually oak—and between the timbers were intertwisted wattles, over which mud was plastered.

This mud or clay, sunbaked only, was in later days painted white, and the timber black. This was the origin of the Cheshire "black and white," "magpie," or as it was sometimes called, "wattle and daub" style of building.

Astbury Church

A few of the churches and some of the noblest and most beautiful of the Cheshire mansions and the most picturesque cottages are in this style. Chester, even the most modern building, is largely black and white, and very "old-world" it looks.

There is little real Saxon work in any Cheshire church : the old wattle and daub Saxon churches have naturally vanished. But the style lasted, and in a few places we find timbered churches which were erected in the thirteenth and fourteenth centuries. The three oldest are at Nether Peover, Warburton, and Baddiley.

Nether Peover Church

There is thirteenth century timbering in the interior of Nether Peover, fourteenth century plaster, and a fine sixteenth century tower. At Warburton the timbers of the church are very old and massive, but it has a brick tower and porch which were built when Charles I was king. When the clay cracked and the wattles perished they were replaced by brick and stone, but these were painted in imitation of the old plaster.   At Marton and Siddington we have old timber and more modern walls.

Norman Doorway, Prestbury

Norman work is still preserved in many of the churches.   A very fine Norman doorway, which was removed from the old church at Prestbury when it was rebuilt in the eighteenth century, is built into the school-house close to the church. In the north aisle of Chester cathedral there is some Norman work; part of the tower of Acton Church is of the same date, and at Bebington and Shotwick in Wirral we can find twelfth century architecture, though at the latter place it is only preserved in the porch. Norman work is also well seen in St John's Church, Chester.

The Early English style is preserved in many churches, but often the original stones have been renewed. Mobberley was first built in this style, but there are later additions; Nether Peover has much Early English work in it, and at Astbury, Acton, Audlem, and Chester Cathedral we may also trace it. Of the Decorated style we see examples at Bebington, Great Budworth, and a few other places, but by far the greatest number of the sandstone churches of Cheshire were built from the end of the fifteenth to the middle of the sixteenth century, when the Perpendicular style obtained. Perhaps the finest example of this style is the church at Nantwich, which has a central octagonal tower 110 feet high with an outer stair turret. The church contains earlier Decorated work, and twenty finely carved canopied oak stalls, which are supposed to have come from Vale Royal Abbey. Other excellent examples of the work of this period are the churches at Acton, Alderley, Astbury, Audlem, Bunbury, Grappenhall, Malpas, Middlewich, Mottram, Over, Shotwick, and most of the cathedral at Chester.

It will be noticed that several churches are examples

of more than one style, which tells of their enlargement or alteration at different times.   Very many were rebuilt or restored in the eighteenth and nineteenth centuries, and often the original work was carefully copied.   Many of these churches, we see from *Domesday Book*, stand on the sites of Saxon churches.

There is one feature of the Cheshire churches that

Nantwich Church

must not be overlooked; many of them contain effigies, frequently in alabaster, as well as memorial tablets of the famous Cheshire men.   The older representations of the knights are in armour, and the workmanship is well worthy of study, as also are the foot-rests, usually the crest, and the pillows upon which the head reclines.

Ruins of the monastic houses which were such a

feature in Cheshire before the Reformation are few. Vale Royal Abbey, which was founded by Edward I in 1277, though now a mansion, has lost all external traces of its former condition, and the earlier monastic house at Darnhall has entirely vanished. Norton Priory, founded by the son of the first Baron of Halton, and afterwards made an Abbey by Henry VI, is also a country seat. Stanlow, where the monks suffered so much from the inroads of the Mersey tides, is only marked by a few stones built into cowsheds, and its grange at Ince by windows in dwelling-houses and farm buildings. At Wincle and a few other places we can trace granges, but really the only building of this character which remains is at Chester Cathedral, where fortunately some of the original rooms of the Abbey are preserved. Combermere Abbey, with whose history the fine church at Acton is intimately connected, is now a residence.

The remains of Birkenhead Priory, which was founded by Hamon de Masci about 1150, are now preserved by the Birkenhead Corporation. The monks, prior to the time of organised Mersey ferries, used to house travellers in the Priory until the weather was calm enough for the crossing, and made a small income by rowing the passengers over.

22. Architecture. (b) Military—Castles.

There were, of course, castles in England before the days of the Normans, and in Cheshire some of the sites of the British and Roman forts retain the name Castle.

There are, too, in Cheshire one or two castles whose history is very recent; we must not suppose that Peckforton and Bolesworth Castles, for instance, were built in ancient days. The Normans are said to have erected more than a thousand castles in England, and some of these guarded the troubled lands of Cheshire against the rebellion of the conquered Saxons and the inroads of the turbulent Welsh. Some of these castles belonged to the Crown, and others were in the hands of the Earl or his Barons, who ruled the county for the king.

The Norman castles were surrounded by very thick walls, and the gateway was guarded by towers or a gatehouse. Without the walls there was generally a moat filled with water, which was crossed by a drawbridge, and the gate itself was supplied with a great iron grid or portcullis which could be lowered and so close the entrance very effectively. Within the area encircled by the walls was a second, and in some cases a third wall or line of defence, and the donjon or keep stood in the centre and was always provided with a well, so that the defenders would be supplied with water in times of siege.

Chester Castle, of importance since Hugh Lupus built it in 1060, stands on the ancient site overlooking the Dee, but it has been rebuilt and is now the depôt of the Cheshire regiment. At Macclesfield, where there was a baronial Norman castle, a few carved stones and a groined archway, behind a modern shop, are all that remain. On the Welsh border, guarding the Dee and the marshes, were Shotwick, Dodleston, Shocklach, and Oldcastle, now only to be traced by the mounds on which they stood or the depressions of their moats, while at Middlewich, Northwich, Frodsham, Malpas, and Ashley, where castles or outlying forts were situated, nothing remains. At Runcorn the very rock on which the castle stood was removed to make room for the Ship Canal.

Beeston Castle

Two ruined castles are still standing. The least destroyed is on a rock 500 feet above sea-level at Beeston, opposite the modern castle of Peckforton. It was built by Ranulph Blundeville, Earl of Chester, about 1220. The moat without the outer wall has vanished, but there is a ditch running in a semicircle round the castle proper, and opening out at either side on the steep crag which was its greatest protection.   A well, 366 feet deep, is bored through the heart of the rock to the level of Beeston Brook.   Halton Castle, built about the same time, is even more of a ruin than Beeston.   Like the latter, it stood on a rocky bluff overlooking a valley and guarded the ford of the Mersey and the narrow gap at Runcorn.   An ancient tower at Doddington is all that is left of the castle which was erected by Sir John Delves in 1364.   Delves was a squire of Lord Audley, who under the Black Prince fought at Poictiers; the battered statues of the Prince, Lord Audley, and his four squires, all Cheshire men, decorated the tower.   At Brimstage in Wirral there is a peel tower attached to a modern farm, which is supposed to be the remains of the castle erected in 1398.   In other places there are names suggesting peel towers, but they themselves have vanished.

23. Architecture. (c) Domestic—Famous Seats, Manor Houses, Farms, Cottages.

Cheshire, ever since the Norman barons made their habitations in it, has been famous for its county families; it has been said that in no other county in England in proportion to its size are there more country seats.   So long as the county was the scene of constant conflict the mansions and even the farms were fortified, and there are moats round many of the older houses.   It is not unusual to find a moat surrounding a modern farm which stands on the site of some almost forgotten hall. By no means all the halls of to-day are occupied by the original families; indeed, many of them have been farms for generations. Most of the present country mansions were built in Tudor times, several being occupied by the agents of Henry VIII, who settled them on the land taken from the despoiled monastic houses after the Dissolution. Some of the halls, however, are older.

"Wattle and daub," originally willows and mud with a timber framework, was the original material of which the houses were built in the greater part of Cheshire. In most cases the plaster has entirely vanished, and the mud walls have been replaced by stone or brick. In the hill country, where the hard grits were handy, stone has been used for building purposes, and one characteristic of these stone houses is that they weather and look old rapidly, so that it is difficult to tell their age. As a rule they are of later construction than the black and white buildings of the lowlands.

Baguley Hall, now a farm, is one of the oldest of the existing halls; it was built during the reign of Edward III, but has been very much altered. Its old hall, open to the roof, remains, but modern rooms and offices have been attached at either side.

In the *Percy Reliques* we read of a Legh of Baguley, whose ancestors

> Had yearded there long,
> Before William Conqueror
> This Cuntry did inhabit.

Old House: Prestbury

Tabley Old Hall, beautifully preserved, dates from 1380, and stands within a wide moat. Chorley Hall, a farm, contains fifteenth century work and later black and white.

Bramhall is perhaps the finest example of black and white work in the country. It has thirteenth and four-

Moreton Old Hall

teenth century wings, and a Tudor great hall. It was originally quadrangular. Harrison Ainsworth introduces Bramhall into *Rookwood*, referring to the old traditional right-of-way through the hall. Bramhall was carefully and faithfully restored when it was in danger of decay, and is now a perfect ancient modern mansion.

The moated quadrangular Moreton Old Hall is, like Bramhall, decorated inside with the Royal arms of Queen Elizabeth as well as the arms of former owners, and the building is a most excellent specimen of sixteenth century black and white domestic architecture.   It contains a tiny chapel and a priest's hiding place, and a secret passage

Moreton Old Hall

led beneath the moat.   Around so ancient a dwelling we should expect to find much history, but really little is known.

Wythenshawe Hall, built during the reigns of Henry VIII and Elizabeth, is a most picturesque building in quite another style.   It was besieged by Col.

Dukinfield and, it is said, by Fairfax during the Parliamentary wars, but an ancestor of the present owner defended the hall successfully ; Tatton of Wythenshawe is an ancient name in Cheshire.

At Dutton Hall, on the Weaver, the old plaster has been replaced by brick and the great hall divided up to provide more rooms, but the porch and carving show what a grand mansion it was in early Tudor days. The Duttons are said to have been descended from Odard who was established here in 1086, and Duttons again and again turn up in Cheshire and indeed in English history. It was a Dutton who fought under the Black Prince at Poictiers, it was a Dutton who gained the thanks of Henry VIII for quelling a rebellion at Norton after the Priory was dissolved.

Poole Hall, a beautiful stone building with many gables, a fine turret, and massive chimney-stacks, was built in 1570 and later. At Huxley there is an ancient hall, now a farm, surrounded by a moat, with a most picturesque archway over the moat bridge ; it is mostly sixteenth century work.

Dorfold Hall, near Nantwich, was built in 1616, by Sir Ralph Wilbraham. Its Jacobean ornamentation within is elaborate, but the real beauty of the building is in its gables, chimney-stacks, and mullioned windows. Not long after it was built it was the scene of many fights, for it threw in its lot with Parliamentary Nantwich when Lord Capel was at Whitchurch, and being outside the defences of Nantwich had to withstand attack again and again.

Harden or Arderne Hall stands in a fine position over-looking the Mersey above Stockport. Ardernes, Ardens, Hardens, or Hawardens, all of one family, occupied many seats in Cheshire and Flintshire, and the Earl of Hadding-ton, their descendant, is hereditary Chief Forester of Delamere to-day. Harden Hall is in ruins, but an

Arley Hall

octagonal turret and other portions remain of the Tudor building, though not of the one in which tradition says John of Gaunt resided. A Stockport mansion of the family is now preserved as a bank. Marple Hall, a very beautiful place, was the home of John Bradshaw, though before his day it had been owned by the Vernons, relatives

of the Vernons of Haddon Hall.   It stands on one of
those naturally defensive sites overlooking the river
Goyt.

It is impossible here even to name all the famous
Cheshire seats which apart from their architecture are
so intimately connected with history.   Such buildings
as Gawsworth, Hollingworth, Lyme, Oulton, Carden,
Broxton, Calveley, Capesthorne, Arley, Utkinton, Burton,
Gayton, and many others, well deserve mention, but
Cheshire is full of these ancient seats, and all of them
have historical associations.   In the old towns of Chester,
Nantwich, Sandbach, Malpas, Tarporley, and Middlewich,
there are many most interesting houses, most of which
were the town houses of the Cheshire gentry.   The
majority of these are built in the Cheshire black and
white, and quaint gable-ends or ancient chimney-stacks
and ornamented rain-spouts catch the eye as one passes
through the towns, carrying us back in imagination to
the days long past, in spite of the plate-glass shop fronts,
trams, or electric-light standards in their immediate
vicinity.   Of newer buildings mention must be made
of Eaton Hall, the seat of the Duke of Westminster, an
enormous mansion near Chester, entirely rebuilt about
1850.

Away from the towns are the many-gabled, timbered
farms, and black and white thatched cottages with their
gardens full of flowers and herbs of old time association.
Within them we shall find stone floors, sometimes sanded
still, old oak furniture, and inhabitants whose speech,
habits, and in many cases courtesy and hospitality are as

Old House, Bridge Street, Chester

little changed from those of their ancestors who built the houses as are the roofs and gables unaltered by time.

Amongst the hills there are collections of cottages called "folds." Many of these have been altered, but some remain unchanged; their doors and lower windows all look inwards to a cobbled yard, and only the windows of the upper rooms show on the outside of their bare time-worn stone walls. They were built round this yard for protection from raiders and cattle-thieves; in time of danger the inhabitants could drive their flocks and herds into the yard, block up the entrance and defend their houses from the upper windows. These hill houses are mostly roofed with flags or shingles—Kerridge flags are largely used to-day for modern houses, and are more attractive looking and serviceable than slates. In the lowlands thatched cottages are abundant.

In rural Cheshire the village pound, in which stray cattle were placed till redeemed, still stands in many places; as do the village stocks in which the human wanderer or village transgressor of the law remained for punishment, the village cross, or the steps on which it stood, and other relics of forgotten rural customs. Indeed in few counties near the manufacturing centres can we find so many relics of the simple life led by our ancestors. It is not remarkable then that Cheshire is a favourite haunt of the artist. Near Alderley there is a lane so full of beautiful old cottages that it is named Artist's Lane, and there are many such lanes further from the thickly populated districts.

24. Communications—Past and Present. Roads, Railways, Canals.

Long ages before cotton, silk, coal, and the chemical trades had come to revolutionise the county, Cheshire was a trade highway. Through the Cheshire forests and across its marshes were the communicating tracks between the villages of the Cornavii, the roads used by the British merchants who trailed their weary slaves from Wales or Ireland, and their loads of skins and even gold from the mountains of Wales, towards the southern lands. One great highway, right across the county, was in existence when the Romans came, and they paved the sandy or muddy track—the forest path became a road.

The Romans were our first road-makers, and carefully constructed their highways so that they might be able to move troops rapidly between their military camps. These military highways were used as trade routes, and the most important one, known as Watling Street, ran from the coast of Kent, through London, and northward through the Midlands. In Cheshire there were two Watling Streets—the one, the road from London to Chester, being Great Watling Street, and the other, from Chester to Manchester, North Watling Street. Great Watling Street entered Cheshire near Oldcastle on the Shropshire border, and passed northwards through Stretton and over King's Marsh to Aldford, where it forded the Dee, passing through what is now Eaton Park to a second ford close under Chester walls.

North Watling Street evidently followed the older British trade route north-east from Chester, and is really a continuation of Great Watling Street, though changing direction towards Manchester. It passed through Barrow and Kelsall, and over the Delamere Hills north of the camp at Kelsborrow and south of that at Eddisbury,

North Watling Street—now the Chester-Manchester Road

running direct to the salt-works at Northwich. From thence it ran past Holford and northward by Dunham and Altrincham to Stretford, or "street ford," where it crossed the Mersey.

Another important Roman road ran almost straight from the coal mines at Wigan in Lancashire to the Mersey ford at Wilderspool, where it passed through the town of Veratinum, and from thence through North-wich to Kinderton at Middlewich. This road between Northwich and Middlewich is now called King Street, and is wonderfully straight, as we can see by looking at the long line of telegraph posts. King Street is merely the altered name Kind Street. From the Roman station at Kinderton it ran south to Newcastle in the Potteries, or really to the Roman station at Chesterton.

There were less important roads connecting Wilders-pool and Chester, Manchester and Macclesfield, Man-chester and Buxton, and one passing from Buxton due north, crossing Cheshire in the east, from Melandra Castle near Glossop to Bucton Castle on the Yorkshire border.

If we look at a Cheshire map we shall find two places, a long distance apart, called Stretton, and in several places the name Pepper Street. Both these names are evidences of the former existence of Roman roads; Stretton is "street town," and Pepper Street, probably the corruption of some older name, is applied to certain roads, but the derivation is not certain. There was a Pepper Gate in the Chester walls, which was built up by a mayor after his daughter had eloped through it, which gave rise to a local proverb similar to shutting the stable door after the horse is stolen.

For many hundred years after the Romans had de-parted, Cheshire roads were poorly cared for, and in consequence we find them, even so late as the end of the eighteenth century, in a very bad condition. Miss Anna Seward, writing in 1794, speaks of the "frightful ruggedness" of the Wirral roads, and she was not the only one who commented on the perilous state of the highways. Scratched on a pane in Shocklach Church is the suggestive remark of a traveller in 1756 —"The roads were so bad we were in Danger of our Lives." Highwaymen and footpads carried on their business on the very outskirts of the towns. Tradition says that Dick Turpin killed a man at Newbridge Hollow, close to Altrincham, but the story of Higgins, the Knutsford highwayman, who kept his hunters and visited the neighbouring mansions as a guest, only to rob the very ladies he had danced with when they were driving home along the lanes, is no tradition, for De Quincey and Mrs Gaskell told his story in fact and fiction. For over thirty years the mouldering bones of James Price, gibbeted in chains for robbing the Warrington mail, swung in the wind on Hoole Heath almost under Chester walls, and our grandfathers, less than an hundred years ago, dare not travel along the country roads without a horse pistol in their holster.

In the middle of the seventeenth century toll-gates were erected on the roads, the money collected in tolls being devoted to the maintenance of the highways, and many of these bars had not been removed fifty years ago, whilst the old toll houses still stand at important crossings.

Most of the great highways from London to the north and north-west pass through Cheshire. The Holyhead Road enters at Bridgemere, and passes through Nantwich to Tarporley, and thence by Tarvin to Chester. A road from London to Manchester crosses the Dane at Holmes Chapel and proceeds via Knutsford to Bucklow Hill, where it joins the Manchester and Chester Road. Another enters near Bosley and takes Macclesfield and Stockport on its way; this is joined by yet another which passed Buxton on its route, and enters Cheshire at Whaley Bridge. From Buxton a road, now much improved in gradients, ran over the moors by the famous "Cat and Fiddle Inn," over 1600 feet above the sea, which disputes with certain other inns the honour of being the highest licensed house in England. This road is continued through Macclesfield to Monks Heath, where it joined the route of the Liverpool mail, which entered Cheshire at Odd Rode, passed through Congleton and Capesthorne, and was continued through Knutsford to Warrington. The London and Warrington road branched at Odd Rode and passed through Sandbach, Middlewich, and Northwich, following the line of the Roman Kind Street; at Comberbach, near Budworth, there is a good example of the old coaching inn—"The Cock," which figures in one of the journeys of "Drunken Barnaby." In addition to these are numerous excellently kept roads connecting all the chief towns in Cheshire, and the cyclist knows that Cheshire roads are as good as any he can find in the country.

In 1772 there was only one daily coach to London from Manchester, the "London Flying Machine," which took two days for the journey. There was a daily "Stage-Coach" to Liverpool, but all other communication through Cheshire from Manchester was by waggon. By 1821 the Cheshire roads were crowded with daily coaches; those from Manchester alone which passed through the county numbered nearly 40. Of these three were Royal Mail coaches to London, Birmingham, and Sheffield, and there were six daily coaches, carrying mails, to Chester, and six to London, in addition to many other "post" coaches. Besides these were numerous coaches from Chester to all parts, and others crossing to Birkenhead for Liverpool, and to Warrington for all northern towns.

The history of internal communication by waterways or canals practically starts from Cheshire, for though there were canals in England at an earlier date, it was not until the Duke of Bridgewater, the father of canal navigation in this country, employed James Brindley to construct the waterway from Worsley to Runcorn in 1759–61 that it was proved that a great canal was of value. This canal begins at Worsley in Lancashire, but it enters Cheshire six miles from its start, and the rest of its course, through Altrincham and Lymm, is in the county. The canal, largely used for the conveyance of cotton from Liverpool to Manchester, is now one of the most profitable branches of the Manchester Ship Canal undertakings. In 1766 Brindley constructed the Grand Trunk Canal, which crosses the centre of Cheshire, passing through Northwich, and the Ellesmere and Chester Canal followed in 1772. The Mersey and Irwell, the Trent and Mersey, the Chester and Nantwich, the Shropshire Union, and the Peak Forest, are the chief canals in Cheshire to-day. They have passed into the hands of the railway companies, but without exception remain busy and economical trade highways, for they are connected with the canal systems all over England. The carriage of goods by water is cheaper than by either road or rail, and now that steam haulage has largely replaced the slower work of horses and mules, the canals of the county are of the greatest value to its commerce. The Manchester Ship Canal demands a separate section.

Before we proceed any further let us think for a

Ship Canal and Mersey, Runcorn

moment why it is that Cheshire has so many old and important roads and canals. One answer explains it all—its position, but the reasons why this position is important are manifold. First then, it is situated in between great commercial centres, thickly populated manufacturing districts, and one of the greatest English ports. Secondly it has natural productions and manufactures of its own which are only to a small degree consumed within its boundaries. Thirdly, and this explains a great deal, its geographical position is peculiar. England, from the north to Derbyshire is practically divided into east and west by the Pennine range of hills—its "backbone," and all the traffic between Scotland and northern England must pass on one side or other of this hill country ; Cheshire lies between the sea and these hills, so that everything which travels by the west coast route, or by the main route through Wales to Ireland, must go through the county.

Now we can go a step further and consider the most rapid and modern means of travelling or conveying goods —the great railway system of England. We have in Cheshire main lines or branch lines of five of the first railway companies ; the London and North Western main line to Scotland passes right through the county from Crewe to Warrington ; its main line to Ireland, *via* Holyhead, from Crewe to Chester ; its lines to Liverpool and Manchester, from Crewe to Warrington, and from Crewe to Stockport. Its system too is provided with numerous branch lines, but the great hub or centre is at Crewe. Crewe is perhaps the most important junction in England ; its station covers an area of over 137 acres, and at it are the locomotive works of the company, employing about 8000 hands. The Stockton and Darlington railway, the first line for passenger traffic, was opened in 1825; it was followed five years later by the Liverpool and Manchester, and in 1837 the first Cheshire railway—the Grand Junction, connecting Birmingham and Liverpool, through Crewe—started the great railway system which is now a network over the country.

The Great Western has a line to Chester and Birkenhead; the Midland crosses from Marple to Stockport; the Great Central has a main line from Woodhead to Godley, and several branch lines; the Great Northern has trains to Manchester over the Great Central and Midland lines; and the Cheshire Lines Committee has a line right through the county from Manchester to Chester, and others connecting Stockport and many towns with the various lines of other companies. Indeed for business purposes all the lines are linked together.

Thus from the time when the British traders drove their Irish slaves through Chester and over the Delamere pass, and the Romans brought Scotch merchandise along Kind Street, until the present day when the Irish Mail and the Scotch Express rush through at over a mile a minute, Cheshire has been on the main highway from the southern shores and London to the north and west.

25. The Manchester Ship Canal.

The Manchester Ship Canal may fairly be called the latest, largest, and most important of the Cheshire waterways. Of its total length of $36\frac{1}{2}$ miles all but eight actually pass through or along the border of Cheshire, and it is the commercial highway for many large and important towns in the county, for it is connected with the Weaver Navigation and the Bridgewater Canal.

The growth of the cotton trade, even early in the last century, suggested to Manchester business men the notion of a canal capable of floating ocean-going vessels as a cheap means of bringing cotton to Manchester and shipping manufactured cloth. In 1825 a scheme was brought forward to cut a canal from Parkgate to Chester and thence through the county to Manchester, but it was not until nearly sixty years later that the increasing railway rates produced any satisfactory plan in the minds of Manchester men. In August, 1885, the Manchester Ship Canal Company was incorporated.

Daniel Adamson, first chairman and leading spirit, hoped that the capital, then estimated at £8,000,000, would be raised by those people who were most likely to benefit from the canal, and it was arranged that single shares of £10 each could be purchased by easy instalments. Later it was evident that this method would not float the enterprise, and the difficulties which were met with made it first necessary to appeal to speculative financiers and later to the Manchester Corporation, which by a loan of £5,000,000, really cleared the obstacles away. In June, 1908, the total expenditure on capital account, which included the sums spent on the purchase of land and the acquisition of the profitable Bridgewater under-taking, stood at £16,698,230, more than double the first estimate.

In November, 1887, the work, with Sir E. Leader Williams as engineer and T. A. Walker as contractor, was begun; two years later Walker died and after his executors had struggled with the difficulties for some time

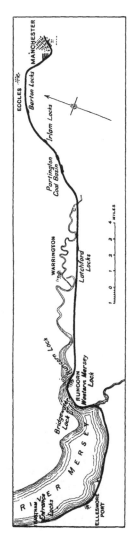

The Manchester Ship Canal

the Company took the contracts over and carried the work through. Legal disputes over the value of land, more than one serious landslip which ruined some of the big steam-navvies and other expensive plant, the under-estimated cost of raising the railways to a sufficient height to safely cross the canal, and other difficulties were only met by the pluck and determination of those interested in the enterprise, which for a long time threatened to end in failure. Time, however, has proved that these men were right, for, except in a year of exceptional trade depression, the profits of the canal have steadily though slowly increased year by year.

About three-and-a-half years after the work began the Eastham, or seaward section, was filled with water, and on July 16th, 1891, a fine flotilla of boats passed through the tidal locks. Two months later the canal was full to Weston Point and communication was opened with the Weaver, at once securing the important salt and chemical traffic of Northwich, Winsford, and the Cheshire salt country. On November 17th, 1893, the water reached Latchford locks, and Warrington docks were ready for business, and a week later, on the 25th, the canal was filled with water from end to end.

An informal opening, when 71 vessels passed into the Manchester docks, took place on January 1st, 1894, and Queen Victoria formally opened the canal on May 21st of that year. In 1905 the present King and Queen declared the fine new dock at Salford open.

The Ship Canal is in several sections, and is provided with five sets of locks. These locks are in pairs, and have double gates so as to save water when small craft only are ascending or descending. The locks at Eastham, through which vessels enter from the channel dredged in the wide Mersey Estuary, are tidal, but can be entered at the lowest state of the tide. The level of the water at Manchester is 70 feet above sea-level or 58 feet 6 inches above low-water in Eastham Pond.

The Weaver Sluices, Manchester Ship Canal

Above Eastham the canal proceeds by a long curve of 13 miles along the Cheshire shore of the Mersey Estuary, passing Ellesmere Port, where by smaller canal and rail there is direct communication with the Staffordshire potteries and iron-works. There is considerable wharfage at Ellesmere Port, and the canal is widened to allow vessels to lie alongside and tranship to flats and

barges without impeding the ordinary waterway. Between
Eastham and Latchford the depth of the canal is 28 feet,
rising to 33 at high tide, and the minimum width at the
bottom is 120 feet. The seaward side is strongly embanked, but in places the course is cut through the
sandstone rock. A little before Weston Point, the port
for the salt and chemical trades, is the junction with the
river Weaver, and the seaward wall of the canal is formed
by ten massive sluices to carry surplus water into the
estuary. From Weston to Runcorn is practically one big
dock, and at Runcorn, where the Ship Canal runs close
alongside the river, it is bounded by a concrete wall which
is 2950 feet long and 16 feet wide at the top. Here the
Bridgewater Canal descends through a series of locks to
the level of the Ship Canal; here too the canal and river
are crossed by a high-level railway bridge and by the fine
modern transporter bridge.

From Runcorn to Latchford, where there are locks,
the course is very straight, and at Warrington, where it is
still on the Cheshire side of the river, there are high-level
and swing road-bridges and two deviated railway viaducts.
Near Lymm the waterway follows the river in a north-
easterly direction, and once or twice cuts across the
boundary into Lancashire. Partington, which is only
six miles from the nearest colliery in the great South
Lancashire coal-field, is a depôt for coal; the Partington
basin covers 6½ acres of water, and is provided with
20 miles of railway siding for coal trains. Six coal-tips,
each capable of loading 160 tons per hour, can keep any
vessels busy which are moored at the wharves. At

Partington the Cheshire Lines railway crosses the canal, and at Warburton a road is carried into Lancashire over a high bridge. The canal leaves Cheshire at Irlam, where in former years the Irwell joined the Mersey, and flows for the remaining few miles of its course to Manchester through Lancashire. In Lancashire there are locks at Irlam, Barton, and Mode Wheel, and about 120 acres of water available for docks at Manchester and Salford, where the dock estate covers over 400 acres. The canal is provided with 133 miles of railway connected with all the great railway systems; it is lighted by electricity, and fitted at the docks and wharves with powerful hydraulic and electric cranes and elevators to facilitate loading and unloading. The company provide tugs to assist large steamers to navigate the canal, and a boat can come up on its own steam in about eight hours; it may safely steam at from six to eight miles an hour.

It has been roughly estimated that the area directly affected by the Ship Canal is inhabited by 10,500,000 people, mostly living in 100 towns of over 10,000 inhabitants each, whilst the population of eleven of these towns is over 100,000. Several of these towns and a fair proportion of the people are in Cheshire.

26. Administration and Divisions— Ancient and Modern.

The history of county government and of more local government still is closely linked with the administration of the land as a whole, and although times have altered since our Saxon forefathers divided up the land into self-governed divisions, there is a gradual growth or evolution which can be traced for over a thousand years.

The government of each county in Saxon days was partly central and partly local. The chief court of Cheshire in early days was the Shire-moot, which was a survival of an even earlier gathering round the "moot-hill" when the people assembled round their sacred hill or grove to administer the law. Representatives of each local district met at the Shire-moot twice a year, under its two chief officers, the Ealdorman and the Sheriff (shire-reeve), the last of whom was appointed by the king.

Cheshire was divided into Hundreds, each of which at one time probably represented one hundred free families, and the hundreds were divided into townships and parishes, as we call them now. At the *gemot*, or assembly which ruled the township, every freeman could attend, and help to make laws for the town and appoint officers to see that they were enforced. The reeve and tithing-man were the officers, and the latter was a constable, something like the policeman of to-day. The reeve was the chairman or president of the court.

The Normans made many alterations in this government, introducing the feudal system, in which the Earl and his Barons really ruled, under the Church nominally, but practically under the King. Cheshire was in Saxon days divided into more hundreds than later in its history, but we must remember that it at first included within its borders land which now belongs to Lancashire and North Wales.

Beside the courts of the shire and hundred, there were courts of the manor, which were of different sizes, and were under the lord of the manor. The lords of the manor had to render service or homage to the king, and in Cheshire to the Earl, as he had the right of ruling "by the sword." The actual sword of Hugh Lupus is preserved in the British Museum; it was his sign of office over the County Palatine, which was in Norman days a little kingdom in itself. The manor courts were known usually as the court-leets, where the lord met his tenants and arranged all matters concerning the manor, such as rights of common, holding of fairs and markets. The court-leet still meets in a few places in Cheshire, but it now has absolutely no power and is only kept up on account of its antiquity. Until quite recent times there were seven hundreds in Cheshire—Wirral, Broxton, Eddisbury, Northwich, Nantwich, Macclesfield, and Bucklow, and for certain reasons these divisions are still used.

The present method of county government is what we must consider now. The Lord-Lieutenant and the High-Sheriff are the chief county officers. The first is usually a nobleman or great landowner, and he is appointed by the Crown; the Duke of Westminster is Lord-Lieutenant of Cheshire. The Sheriff is chosen every year on "the morrow of St Martin's Day," November 12th. The chief county business is transacted by the County Council, which meets at the County Town, Chester. The Council consists of 19 Aldermen, 59 Councillors and a Clerk. The former are co-opted, the latter elected to their office. The keeping of roads and bridges in good repair, the appointment of police, the regulations with regard to lunatic asylums and industrial homes, and the carrying out and enforcing of laws passed by Parliament are part of the important work of the Council.

The County Council is the central government, but since 1894 new names have been given to the local governing bodies, and they are now called District Councils. They were originally called local boards, and vestries. In Cheshire there are 33 Urban, 13 Rural, and 464 Parish Councils. The large towns, which are called Boroughs, have a different form of government—a Town Council or Corporation, which consists of a Mayor, Aldermen, and Councillors. Three of the boroughs, Chester, Birkenhead, and Stockport, are County Boroughs, having the powers of a County Council. The others are —Congleton, Crewe, Dukinfield, Hyde, Macclesfield, and Stalybridge, whilst parts of the boroughs of Ashton-under-Lyne and Warrington are in Cheshire.

Before we go any further it must be explained that a county may now be considered in three lights, as an Ancient or Geographical, an Administrative, or a Registration County. The borders of all three are different in Cheshire, but we need not trouble about the last, which is purely arranged for the convenience of the registration officials. Up to this point we have considered only the Geographical County, which is enclosed within the old county boundaries. The Administrative County gives up certain portions to other counties and receives others from them, mainly in the neighbourhood of large towns.   At the last census there were 1520 inhabitants of the Administrative County of Derby residing in the Ancient County of Chester ; Lancashire inhabitants were 39 of Ashton-under-Lyne, 2047 of Mossley, and 7983 of Warrington ; and 38 inhabitants of the Administrative County of Shropshire live in Cheshire at Tittenley.

At Threapwood 106 persons in Administrative Cheshire live in the Ancient County of Flint ; at Dukinfield 45, Stalybridge 7402, and at Stockport 17,066 are in the Ancient County of Lancaster.   It will thus be seen that Cheshire gains most in the borders of its boroughs of Stockport and Stalybridge, and loses at Warrington— which is a County Borough—and Mossley.

The total population of the Administrative County with the boroughs is 827,191, and the area 654,825 acres, as against a population in the Ancient County of 815,099, and an area of 657,783 acres.

Cheshire has 11 Poor Law Unions, over each of which is a Board of Guardians, whose duty it is to manage the workhouses and appoint officers to carry on the work of relieving the poor and aged.

For the administration of justice, Cheshire county has one Court of Quarter Sessions, and 14 Petty Sessional Divisions, each having magistrates to try minor cases and award punishments.   Six boroughs have each separate Commissions of the Peace—Congleton, Crewe, Hyde, Macclesfield, Stalybridge, and Stockport.   Chester and Birkenhead have separate Courts of Quarter Sessions and Commissions of the Peace.

Ecclesiastical or Church government has altered little during the last thousand years; the Church existed before the State and had its divisions or dioceses, and its own courts. The dioceses in the south of England are under the care of the Archbishop of Canterbury, and those in the north, Chester amongst them, under the Archbishop of York. The diocese of Chester has a bishop, two archdeaconries, rural deaneries, and parishes. There are 268 ecclesiastical parishes in the county, some of which include two or more civil parishes. Two of the ecclesiastical parishes are in the diocese of Manchester, part of Threapwood in the diocese of St Asaph, and part of Whitchurch in that of Lichfield.

The education of the county is governed by the Education Committee of the County Council, and separate committees for the larger towns and boroughs.

Cheshire is represented in the House of Commons by 13 members of Parliament. Stockport sends two members; Chester, Birkenhead, and Stalybridge each one, and the county is divided into eight divisions—Altrincham, Crewe, Eddisbury, Hyde, Knutsford, Macclesfield, Northwich, and Wirral, each with one member.

27. The Roll of Honour of the County.

When reviewing the history of the county of Chester we had occasion to refer more than once to its many famous men or worthies, and it is well for us to remember those whose names are household words who are associated with our county. Cheshire has a very long roll of honour, and it will be impossible even to mention many of those who have added to its prosperity or glory, or who have by pen or sword or in other fashion helped to build up and rule the land we live in.

Dean Stanley

Cheshire has not actually been the birthplace of any of our kings or queens, but John of Gaunt, "time honour'd Lancaster," the father of Henry IV, was baron of Halton. The two daughters of James II, afterwards Queen Mary and Queen Anne, were, through their mother Ann Hyde, of the family of Hyde of Hyde. Nearly all our kings and queens visited Cheshire, some of them at the head of armies and some in Royal procession or for the sake of sport and recreation.

In the roll of honour connected with the Church are several well-known Bishops of Chester, and three others, associated with other dioceses, who spent their early life in the county. Dr Thomas Wilson, afterwards a famous Bishop of Sodor and Man, was born in a cottage at Burton in 1663. Bishop Stanley of Norwich, born in 1779, was of the Alderley branch of the family, and for over 30 years was rector of the old church at Alderley, but his son Arthur Penrhyn Stanley, the great broad-church mid-nineteenth century Dean of Westminster, bore a more widely-known name. Dr Reginald Heber, poet and hymn writer, who became Bishop of Calcutta, was born at Malpas in 1783; and is perhaps best known as author of the familiar missionary hymn, *From Greenland's icy mountains.* Four Puritan divines, Henry Newcome of Goostrey and Gawsworth, Adam Mantindale of Rostherne, Edward Burghall of Acton, and Samuel Clarke of Shotwick, are intimately connected through their writings with the history of puritanism in Cheshire. When considering the names of those made worthy by their connexion with religion, we must not forget to honour two Cheshire martyrs who died for conscience sake. George Marsh was burnt at Chester during the Maryan persecution, and Margaret Ward of Congleton suffered for her fidelity to Roman Catholicism in the Elizabethan retaliation.

Cheshire has been the home of many great lawyers and statesmen. Thomas Egerton, Lord Chancellor Ellesmere, ancestor of the Dukes of Bridgewater, was born at Ridley in 1540 and was buried at Dodleston; we read of him in connexion with the trial of the Earl and Countess of Somerset for the poisoning of Sir Thomas Overbury. Chief Justice Crewe, who built Crewe Hall, and Sir Humphrey Davenport of Bramhall, were famous men in Stuart days. John Bradshaw, the regicide, was born at Malpas Hall and educated at Bunbury. A Bunbury man, Sir John Williams, became Attorney-General, and made his name by his excellent defence as junior counsel of the unfortunate Queen Caroline.

In the old Cheshire churches are the effigies and memorials of many Cheshire men who fought for king and country. We may recall the valour of Sir Thomas Danyers at Crecy, and Sir John Delves at Poictiers, of Sir Hugh Calveley, "a man of teeth and hands," at Auray and Navarete, of Sir Ralph Egerton and Sir Edward Stanley (immortalised of Scott) at the Battle of the Spurs and Flodden.

We may remember too, Sir George Beeston who commanded the "Dreadnought" in the first line against the Armada, and the rival Royalist and Parliamentarian leaders, Sir Thomas Aston of Aston, and Sir William Brereton of Handforth. Lord Clive was educated in Cheshire and spent his boyhood at Rudheath, and Field-Marshal Viscount Combermere, who took Bhurtpore in 1825, was a Cheshire man. General Sir William

Congreve of Burton attained distinction as the inventor of the war rocket.

Perhaps the earliest Cheshire historian was Henry Bradshaw, a Chester monk who wrote the *Life of*

John Gerard

St Werburgh in the fifteenth century. John Speed, whose sixteenth century maps and histories are famous, was born at Farndon on the Dee. In the seventeenth century Sir Peter Leycester of Tabley and Randle Holme of Chester added valuable literary works to local and general history, and Daniel King's *Vale Royall* was the backbone of Dr Ormerod's *History of Cheshire*, written early in the last century, and still the standard history of the county.

Mrs Gaskell

Geoffrey Whitney, author of the *Choice of Emblemes*, printed at Leyden in 1586, was a Nantwich man and a Cheshire poet, and about the same time flourished John Gerard, another native of Nantwich, whose *Herball or Generall Historie of Plantes* is one of the quaintest and most interesting works of the early botanists.

William Broome, a minor poet, was born at Haslington; he was associated with Pope in his work on the *Odyssey*, and was later lampooned in the *Dunciad*. The third Lord de Tabley, who died in 1894, was at once poet and botanist; his *Flora of Cheshire* is a valuable addition to science.

Mrs Gaskell spent many years of her early life at Knutsford, and has immortalised the place in her tale *Cranford*; she is buried in the graveyard of the ancient Unitarian chapel at Knutsford. Charles Lutwidge Dodgson, best known as "Lewis Carroll," was born at Daresbury, whence perhaps he got his idea of the Cheshire cat in *Alice in Wonderland*. Charles Kingsley wrote some of his tales and poems when Canon of Chester, but his name is more famous in the city in connexion with the natural history society which he founded, and which now has its head-quarters in the Grosvenor Museum.

Whether Huxley can be claimed as a Cheshire man is uncertain. The Huxleys of Huxley were a famous Cheshire family, and the great naturalist was possibly a member of it. He himself, however, declared—"My own genealogical inquiries have taken me so far back that I confess the later stages do not interest me."

28. THE CHIEF TOWNS AND VILLAGES OF CHESHIRE.

(The figures in brackets after each name give the population in 1901, and those at the end of the sections give the references to the text.)

Acton (1920) is a village about a mile north-west of Nantwich, with an old and interesting church, which together with Dorfold Hall, built in 1616, was garrisoned by the Parliamentarian troops in 1643. (pp. 134, 140, 150, 176.)

Alderley Edge (2856) is an urban district, taking its name from the old village of Alderley. It was formerly known as Chorley. Chorley Hall, now a farm, is a beautiful old building, partly built of stone, but with a "magpie" wing, and a moat bridge. Alderley Park, the seat of Lord Stanley of Alderley, is a modern building; the old hall was burnt down. The crest of the Stanleys, to be seen on the park gates, is an eagle with a swaddled child; tradition tells that an ancestor of the family was carried away by an eagle which fed the babe instead of feeding on it. (pp. 13, 33, 38, 41, 98, 120, 121, 127, 140, 148, 176.)

Alsager (2597) is a large urban district, five miles south of Sandbach, close to the Staffordshire border.

Altrincham (16,831) is an important market-town, eight miles south of Manchester. De Quincey gives an amusing account of its market in his youthful days—"the gayest scene I ever beheld." The town still has a mayor and court-leet, which, however, has no real power. Broadheath is really the manufacturing portion of Altrincham. (pp. 89, 156.)

Ashton-on-Mersey and **Sale** (17,651) are large residential districts with a station on the Manchester and Altrincham railway. They are rapidly growing, and the combined population is now much greater than in 1901. There was an extensive moor at Sale but it has been entirely built over. (p. 115.)

Audlem (1455), six miles south of Nantwich, is an ancient market-town. The old open-air market shed stands below the

Anderton Lift

substantial square-towered church. At the seventeenth century free grammar school many of Cheshire's worthies were educated. (pp. 21, 68, 140.)

Barnton (2792) is a busy village on the high ground above the Weaver on the outskirts of Northwich. At Anderton there is a famous hydraulic lift for raising canal boats from the Weaver to the North Staffordshire Canal; the boats are raised or lowered in tanks or docks. The lift is an almost unique example of canal engineering.

Bebington, Higher (1540), is really a residential outskirt of Birkenhead. The far-famed Storeton Quarries are close to the village. (p. 35.)

Bebington, Lower (8398), is now an urban district including the model village of Port Sunlight. Nathaniel Hawthorne says of the church at Lower Bebington that it "looks old, old, old," and so it is, for some of the original Norman pillars are excellently preserved. (pp. 60, 92, 140.)

Birkenhead (110,915), a county borough, is the largest and most important town in Cheshire, and is connected with Liverpool by ferries and a tunnel under the river. Birkenhead has many fine modern buildings and the remains of its ancient priory; on its outskirts are extensive residential districts, largely occupied by Liverpool men. Its docks are under the management of the Mersey Dock and Harbour Board. (pp. 18, 38, 46, 59, 82, 88, 90, 92, 100, 102, 106, 107, 142, 172, 173, 174.)

Bollington (5245) is a stone-built town three miles north of Macclesfield; it is largely engaged in the cotton trade. It is almost surrounded by hills, and is overlooked by "White Nancy," a small white building at the north end of Kerridge ridge, which is visible from many miles away. There are extensive freestone quarries at Kerridge. (pp. 21, 32.)

Bowdon, Dunham Massey and **Hale** (9994) are three urban districts which are closely connected, and the population of the last of the three is rapidly increasing. The parish church at Bowdon stands on the summit of a great sand heap, which originally stood above the surrounding marshes; the name means the hill in the bog. Dunham Park is the seat of the Earl of Stamford; the Hall was rebuilt in the eighteenth century, on the site of older halls and a Norman castle. Hale Moss has been drained and built upon, and a portion is converted into a public park for Altrincham. (pp. 21, 41.)

Bredbury and **Romiley** (7107) are busy villages engaged in the cotton trade. The ancient chapel of Chadkirk, formerly in the hands of the Nonconformists, fell into ruins at the end of the seventeenth century, but has been restored; it is a Saxon foundation. (pp. 88, 90.)

Bromborough (1891) is a quiet village in Wirral overlooking the Mersey; it is now an urban district. (pp. 60, 109.)

Buglawton (1452) is now practically part of Congleton. On Congleton Edge, above Buglawton, is the ancient stone circle known as the Bride Stones. (p. 120.)

Bunbury (820) about four miles south-east from Tarporley, is a quiet country village with an ancient Decorated and Perpendicular church containing several interesting memorials and effigies. Beeston Castle, the finest ruin in Cheshire, and Peckforton Castle, the seat of Lord Tollemache, built in 1844–50, are in the parish of Bunbury. (pp. 38, 41, 115, 140, 144, 177.)

Cheadle and **Gatley** (10,820), connected by tramlines with Stockport, are growing residential districts. There is an old church at Cheadle, and several old halls in the vicinity.

Chester (38,309), a county borough, is the county town of Cheshire. It has a fine and ancient cathedral, several interesting old churches, and a castle, now the depôt of the Cheshire regiment, and is surrounded by walls, which in many places have Roman foundations. The latest and best length of Roman wall to be exposed was discovered during the erection of the new telephone buildings; it has been carefully preserved. Chester is full of Roman remains and mediaeval buildings. The streets are in many places lined by "rows," which are very old; there are two levels of shops, a footway is constructed over the lower shops, so that there is a covered pathway in front of the upper shops in addition to the ordinary sidewalk. Bishop Lloyd's Palace, God's

The Rows, Chester

The Rows, Chester

Providence House, the Stanley Palace, and the Yacht Inn are amongst the old houses of Chester. Most of old Chester was built in black and white, and the ancient character has been preserved in its modern buildings. The Dee is crossed at Chester by the Dee Bridge (the ancient bridge), by the Grosvenor Bridge opened by our late Queen, a suspension bridge, and the railway bridge. The Roodee is a fine open space many acres in extent,

The Town Hall, Chester

bounded by the walls and the river; the Chester races are held here. The stump of a cross, said to have been the Rood which gave origin to the name Rood Eye, stands in this open space. Handbridge, where Chester fishermen live, is that portion of Chester on the Welsh side of the river. Eaton Hall, the seat of the Duke of Westminster, stands in a well-wooded and beautiful park, on the river above Chester. The Technical Schools and

Chester Wall and the Roodee

the Grosvenor Museum, containing excellent natural history and archaeological collections, are close to the Castle.   (pp. 2, 3, 9, 23, 24, 32, 51, 52, 80, 81, 90, 92, 100, 105, 107, 109, 111, 112, 114, 115, 116, 123, 125, 127, 134, 140, 143, 152, 172, 173, 174, 176.)

Congleton (10,707) is a municipal borough engaged in the silk trade and other industrial occupations.   It contains some old and many good new buildings.   On the Dane near Congleton is a deserted village named Havannah, its ruined silk mill having been built about 1761 when the capital of Cuba was taken.   A cigar factory was started in this village, but the rows of cottages still stand empty and in ruins.   (pp. 23, 31, 89, 90, 172, 173, 176.)

Crewe (42,074), the great junction of the London and North Western Railway, is a flourishing municipal borough, but it is really not Crewe at all but Monks Coppenhall; the population of the civil parish of Crewe is only 406.   The majority of the inhabitants of Crewe are in the employ of the railway company. Crewe Hall stands in a large park near the town; the Hall was almost destroyed by fire about 40 years ago.   (pp. 162, 172, 173.)

Disley (2827), eight miles south-east of Stockport, is situated amongst the hills; it is a growing village.   Lyme Park, the seat of Lord Newton, is an extensive demesne.   The Hall, an old building of great historical interest, stands in a sheltered hollow in the park.   Above it, and a prominent object in the Cheshire landscape, is the square hunting-tower, Lyme Cage.   There are red and fallow deer in the park and on the moor, where in former days the white cattle of Lyme roamed at will.   The Lyme mastiffs are a famous breed of dogs.   (pp. 47, 152.)

Dukinfield (18,929) is a busy municipal borough, and one of the chief cotton towns; it is connected by a bridge with the Lancashire town of Ashton-under-Lyne, a small part of which is within Cheshire.   The old Hall of Dukinfield is now converted into cottages and the remains of the fourteenth century chapel are a portion of a modern Nonconformist building.   (pp. 82, 88, 172, 173.)

Lyme Hall and Park

Eastham (913) is a quiet little village in Wirral, close to the entrance of the Ship Canal into the Mersey.   (pp. 61, 167.)

Ellesmere Port (4082) was formerly known as Whitby Locks, and stands at the junction of the Ellesmere and Shropshire Union Canal with the Ship Canal.   It is now a port of some

Gawsworth Church

importance, being a great depôt for iron ore for the Staffordshire iron-works.   (pp. 63, 64, 92, 108, 167.)

Farndon (564) is a little village connected by an old bridge with the Welsh town of Holt.   Strawberries are largely grown in the neighbourhood of Farndon.   King's Marsh, close to the village, was one of the three Cheshire sanctuaries where any foreigner who sought the protection of the Earl of Chester was allowed to "squat" or reside for a year and a day in a tent or hut, provided he put up no durable building.   (pp. 23, 84, 178.)

Frodsham and **Frodsham Lordship** (4131) are situated on the edge of the Mersey marshes near the mouth of the Weaver, which is crossed at Frodsham Bridge.   The parish church stands on the hill at Overton, which was at one time the village of Frodsham.   (pp. 33, 38, 95, 144.)

Gawsworth (573), three miles south of Macclesfield, has an exceedingly interesting historical church, an old and a new Hall and a beautiful black and white vicarage.   "Lord Flame," a local wit, player, and dancing master, who died in 1773, was buried in a wood at Gawsworth.   His grave, locally known as "Maggoty Johnson's grave," stands amongst the trees near the New Hall.   There is a tilting-ground at the Old Hall, which belonged to the famous Fitton family.   (pp. 152, 176.)

Godley (1691) is a village near Hyde; the Maypole Dairy Co. employ many of the inhabitants in their margarine works.   (p. 92.)

Grappenhall (987) is a village three miles south-east of Warrington.   It has a fine sixteenth century church, and outside the churchyard wall the village stocks are still standing.   (p. 140.)

Great Boughton (1034) is a residential village on the outskirts of Chester.

Halton (1238) stands on the high ground overlooking the Mersey between Runcorn and Frodsham.   The ruins of the castle crest the hill, and close to them is the Chesshyre library, founded by Sir John Chesshyre in 1733, which contains many interesting old volumes—400 in all.   The Royal Castle of Halton guarded the road from Chester to the north at what was once a dangerous place.   Piers Plowman remarks that through

> "the pas of Haulton
> Poverte might passe whith oute peril of robbynge."

Norton Priory, the seat of Sir Richard Brooke, stands in beautiful woodlands overlooked by the rock on which Halton Castle stands. (pp. 142, 145, 175.)

Handforth (911), six miles south of Stockport, is a village and residential district. The Hall, still in good preservation, was the seat of the Breretons; the date over the Brereton crest, a bear's head, muzzled, is 1562. (p. 177.)

Haslington (1791) is a growing industrial village close to Crewe. (p. 180.)

Hazel Grove and **Bramhall** (9791) is a rapidly-growing urban district, lying to the south of Stockport. The old name of the former place was Bullock's Smithy, a name which dates from 1560. The Hall at Bramhall is perhaps the finest and best preserved black and white building in the county, if not in England. (p. 148.)

Helsby (1572) stands at the foot of the steep crag of Helsby Hill, about eight miles north-east of Chester. It has grown in importance since the cable-works were started. (pp. 33, 38, 41, 70, 90.)

Heswall cum Oldfield (2167) is a residential village on the river Dee. Gayton Hall, an ancient seat of the Gleggs, is at Heswall. (pp. 56, 100, 102.)

Hollingworth (2447) is a stone-built town in Longdendale; the inhabitants are engaged in the cotton trade. (p. 152.)

Hoole (5341) is a thickly-populated residential suburb of Chester. Hoole Heath, which was one of the three sanctuaries, was twice the site of battles, first when Malcolm, king of Scotland, invaded Cheshire, and secondly when the Royalist forces were defeated beneath the very walls of the city. Charles I is said to have watched the progress of this battle from the Phoenix Tower on Chester walls. (p. 158.)

Hoylake and **West Kirby** (10,911) are growing water-ing-places in north-west Wirral. Hoylake is a modern town, so also is most of West Kirby, but there is an ancient village round

C. C.   13

the parish church on the hill above the town.　(pp. 18, 52, 56, 57, 71, 100, 101, 102, 104, 111.)

Hyde (32,766), a municipal borough, is the centre of the Cheshire hat trade, and there are many cotton mills in the town and coal mines in the neighbourhood.　Part of Hyde is called Flowery Field, but the industries of the busy town have long since destroyed flowers in this neighbourhood.　(pp. 82, 88, 90, 172, 173, 176.)

Knutsford

Kelsall (670) is a quiet little village on the pass leading over the Delamere hills.　Close to it is the camp at Kelsborrow, and a little beyond it the remains of the once important fortress of Eddisbury.　(pp. 111, 123, 156.)

Kingsley (1066) is a village in the heart of what was once forest land, associated with the name of the Kingsleys, who shared with the Dones the forestership of the ancient forests of Mara and Mondrem, afterwards Delamere.

Knutsford (5172) is a town of considerable importance nearly in the centre of the county. The county gaol and workhouse are at Knutsford. Tatton Hall, the seat of Lord Egerton, stands in the largest park in the county, where there are herds of red and fallow deer and much fine old timber. Tabley Hall, one of the most interesting of the old Cheshire halls, is in the neighbourhood. (pp. 47, 112, 148, 180.)

Lymm Market Cross

Lostock (1640), really Lostock Gralam, is in the salt country just outside Northwich; Brunner, Mond and Co. have works here. (p. 95.)

Lymm (4707) was originally a quiet village where the inhabitants were mostly engaged in fustian cutting, but it is now a residential district for Manchester and Warrington people. Lymm market-cross is one of the best examples of village crosses now existing. It has been kept in repair, but most of it is old;

13—2

it stands on a flight of worn sandstone steps, and at the foot are the village stocks. (pp. 89, 121, 168.)

Macclesfield (34,624), a municipal borough, is the centre of the Cheshire silk-trade. It is a town built on a hill, and the old church stands splendidly overlooking the lower part of the town, the shortest way to which is down a long flight of stone steps. Smith, in the seventeenth century, described it as "one of the fairest towns in Cheshire," and it has kept up its reputation in many ways. It is a very ancient borough, and at one time had the right of punishing its offenders against the law. It has its Gallows Field and Cookstool Hill, and its stocks and scold's bridle are still preserved. (pp. 17, 20, 32, 82, 89, 90, 143, 172, 173.)

Malpas (1139) is an ancient town in the south of the county, but it is now only a civil parish. Its church is a fine building in Decorated and Perpendicular styles. An hundred years ago not a single turnpike road led to the isolated town; nevertheless the town was known to Giraldus Cambrensis, the twelfth century historian, for he made a pun on the name of the bad pass. (pp. 94, 134, 140, 144, 152, 176.)

Marple (5595) is an urban district four miles east of Stockport. Although a cotton town, it is situated in romantic scenery at the foot of the hill country. In the neighbourhood are the ruins of some of the earliest cotton mills. Marple Hall, a very beautiful building, was the home of the Bradshaws, and is still in the possession of their descendants. Near Marple is the so-called Roman Bridge, which though very old is much later than Roman. (pp. 88, 151.)

Middlewich (4669) is busily engaged in the salt and chemical trades, and is an important town. Dr Johnson described it as "a mean old town, without any manufactures, but, I think, a corporation." It is now an urban district and not a borough, but there is plenty of trade in the town.   (pp. 22, 23, 37, 94, 95, 109, 125, 140, 144, 152, 157.)

Mobberley (1353) is a quiet village with an ancient church. A silk-crape mill at one time found occupation for many of the inhabitants, but it could not continue when the trade in crape went out. (p. 140.)

Mottram in Longdendale (3128) is now an urban district in the cotton area; its grey-stone fifteenth century church stands on a hill from which one of the most imposing moorland views in the county can be obtained. (p. 140.)

Nantwich (17,611) is a quaint old town, with many ancient buildings, which was formerly one of the centres of the salt trade. There are medicinal brine baths in Nantwich now. Nantwich has been twice almost burnt to the ground, and it suffered much during the prolonged siege in the Civil War; it was the head-quarters of the Parliament when Lord Capel was holding Whit-church. Welsh Row is an irregular old street which gained its name when the Welsh were trading with Nantwich, buying salt from the springs. It escaped the fire in 1583. Sweet Briar Hall and one or two other old buildings escaped, and some of the houses erected immediately after the fire are still standing. The church is a large and beautiful building, both inside and out. At the end of the eighteenth century there were important cotton mills in Nantwich, but now the trade of the town is mostly in boots and shoes, clothing, and iron; there are also tanneries and corn mills. (pp. 37, 90, 92, 94, 114, 115, 125, 140, 152, 179.)

Neston and **Parkgate** (4154) form an urban district on the estuary of the Dee. Neston is an old town, and Parkgate was first a port and then a fashionable watering-place; it is now becoming a residential district. There is a small fishing popula-tion in Parkgate. (pp. 32, 55, 98, 100, 102, 104.)

Newton (7407) is a civil parish close to Hyde, and is in the cotton district.

Northenden (2127), a village on the Mersey between Sale and Stretford, is growing owing to the workers in Manchester wishing to reside outside the town. It has a fine church. The old halls of Wythenshawe and Baguley are in the neighbourhood. (pp. 146, ℓ149.)

Witton Church, Northwich

Northwich (17,611) is the centre of the salt and chemical trades. It has suffered considerably from subsidences over the brine deposits, but the more recent buildings are put up in wood frames, so that they can be raised whenever their foundations give way. There are deep rock-salt mines, and a large number of white salt manufactories. Small coasting craft can ascend the Weaver to Northwich. (pp. 22, 23, 37, 81, 90, 92, 94, 95, 96, 108, 109, 115, 125, 144, 156.)

Odd Rode (3187) lies south of Congleton. Astbury Church and Moreton Old Hall are in the neighbourhood. (pp. 134, 140, 149, 159.)

Over (7063) is a small thickly-populated town, closely connected with Winsford and Wharton in the salt and chemical trades. Over Church, which stands in the fields some little distance from the town, is an old building, restored in 1543. There are large and growing flashes on the Weaver above Over. (p. 140.)

Prestbury Church

Poynton and **Worth** (2544) lie between Stockport and Macclesfield. Coal is worked at Poynton. Adlington Hall and other old buildings are near the village. (p. 32.)

Prestbury (291) is a small but ancient town. Near its old church are many quaint and beautiful old houses, and there are others outside the town. The Old Vicarage is one of the best "magpie" houses. (pp. 20, 139.)

Runcorn (16,491), now a port on the Ship Canal, has been a busy place ever since the construction of the Bridgewater Canal, which here entered the Mersey by a long series of locks. It is now connected with Widnes by two bridges, one for the railway and foot passengers, and the other the fine transporter bridge, which carries a section of roadway across the Canal and river. (pp. 20, 33, 36, 64, 92, 95, 111, 144, 168.)

Sandbach (5558), at once an ancient and a busy manufacturing town, is in the salt country. Its trade is in salt, chemicals, fustians, and boots. It has an old church and many picturesque buildings, but the oldest and most interesting relics in Sandbach are its Saxon crosses. Wheelock, a busy salt village, is close to Sandbach. (pp. 89, 95, 115, 127, 152.)

Stalybridge (27,673) is a municipal borough, at one time the centre of the woollen trade of Cheshire, but now in the heart of the cotton district. Staly Brushes, in the Tame valley, was a wild uncultivated district, for long the haunt of Manchester and Lancashire naturalists. (pp. 82, 88, 123, 172, 173, 174.)

Stockport (78,897) is a county borough which includes within its boundaries the Lancashire district of Heaton Norris. It is the chief cotton town in Cheshire. It is a growing and busy town, connected by rail and tram with Manchester. Stockport Sunday School, at which all denominations meet freely, is far famed, and was one of the first started when the Sunday School movement began. Stockport suffered considerably in the Luddite riots when the people rose against the introduction of machinery. (pp. 20, 80, 82, 87, 88, 90, 92, 95, 121, 172, 173, 174.)

Stockton Heath (2543) is practically the Cheshire part of Warrington, which is connected with it by many bridges over the river and Ship Canal. Brewing is one of the chief occupations and there are also tanneries. At Wilderspool there is a small street called Roman Road, which stands on the site of the Roman town of Veratinum. (pp. 81, 90, 92, 123, 125.)

Sutton (1294) is a populous civil parish on the outskirts of Macclesfield.

Tarporley (2644), an urban district, is an ancient borough which had a mayor in 1297. The old Hall of Utkinton, the manor house in Tarporley, and the Swan Inn, head-quarters of the Cheshire Hunt, are worthy of note. Eaton Hall, the seat of the Earl of Haddington, hereditary chief-forester of Delamere, is in the district. (pp. 49, 121, 152.)

Tarvin (1093) is an old town with an interesting church, on the high road from Chester to Manchester. There are some historical old homes in the vicinity. (p. 158.)

Timperley (3215) is a civil parish just outside Altrincham. On the Bridgewater Canal, below Timperley, and close to Altrincham, has recently grown the busy manufacturing colony of Broadheath, where engineering and other works have collected round them a dense population. (p. 90.)

Tintwistle (1622) is the highest village of importance in the Longdendale Valley; it is now a cotton town, but includes within its township some 10,000 acres of barren moorland.

Upton by Chester (1769) is a residential village, largely occupied by Chester men.

Wallasey (53,579) is a large and important urban district north of Birkenhead; it includes Seacombe, Poulton, and Liscard. New Brighton, at the mouth of the Mersey, is a pleasure resort and residential district. The last extensive stretch of Cheshire sandhills remains in the neighbourhood of Wallasey. (pp. 17, 59, 68, 71, 73, 100, 106.)

Weaverham (1882), in spite of the fact that it is only four miles north-west of Northwich, is still an old-world village. Salt was worked here in Saxon days, and later the abbot of Vale Royal had a private court and prison in the town.

Dean Row Chapel, Wilmslow

Weston (2115) is a busy port for salt and chemicals at the junction of the Weaver and Ship Canal; there are large chemical works in the town.   (pp. 36, 64, 70, 73, 95, 108, 168.)

Wharton (3319) is a busy town in the salt district. Bostock, a model village near Wharton, has an oak which was planted to replace an ancient one, which is said to be the centre of Cheshire. (pp. 22, 94.)

Willaston (2400) is closely connected in trade with Nantwich; it lies to the east of that town.

Wilmslow (7361), an urban district, was at one time self-contained, but now is one of the residential outskirts of Manchester. Lindow Common, now taken over by the Wilmslow council, is one of the last of the Cheshire mosses. Fulshaw Hall and Hawthorn Hall are interesting old houses near Wilmslow, and the church is old. (pp. 17, 21, 122.)

Winnington (1604) is a busy parish on the outskirts of Northwich with large chemical works. Winnington Hall, once a residence of the Stanleys, is well preserved, and stands in a public park. (p. 115.)

Winsford (10,382), an urban district, is the centre of the southern portion of the salt country; half a million tons of salt are annually shipped from Winsford. The town hall is a modern building in imitation of the Cheshire "wattle and daub." (pp. 22, 92, 94, 108.)

Yeardsley cum Whaley (1487) is an urban district, better known as Whaley Bridge. It stands on the Goyt on the Derbyshire border. Cotton manufacture is the industry of Whaley. The walk up the Goyt Valley, from Whaley Bridge past Taxal and the Powder Mills to the moors near the Cat and Fiddle Inn is one of the most beautiful walks in Cheshire. (p. 159.)

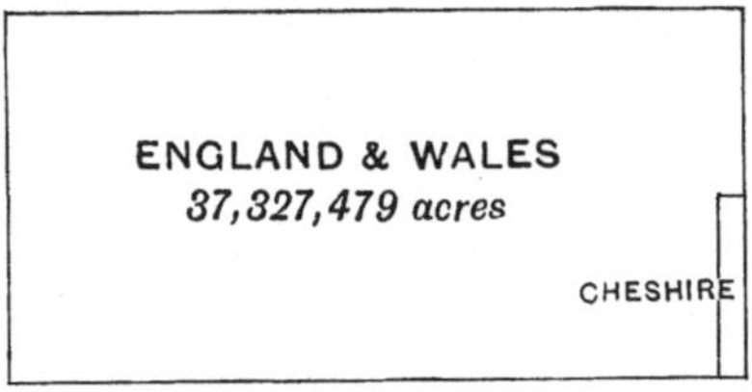

Fig. 1. The Area of the Administrative County of Cheshire (654,825 acres) compared with that of England and Wales

ENGLAND & WALES
32,527,843

CHESHIRE

Fig. 2. The Population of the Administrative County of Cheshire (827,191) compared with that of England and Wales

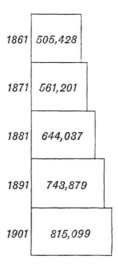

Fig. 3. Diagram showing the Increase in the Population
of Cheshire from 1861 to 1901

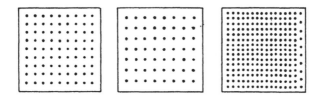

Cheshire, 808   England & Wales, 558   Lancashire, 2347

Fig. 4. Comparative Density of the Population to
the Square Mile in 1901

(*Note, each dot represents* 10 *persons*)

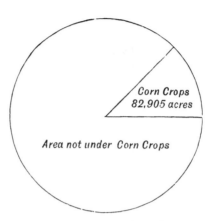

Fig. 5. Proportionate Area of Corn Crops in Cheshire
to Total Area of County in 1907

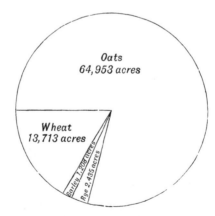

Fig. 6. Proportionate Area of Oats, Wheat, Rye and
Barley in Cheshire in 1907

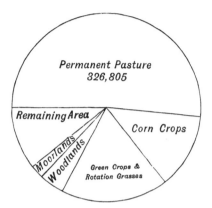

Fig. 7. Proportion of Permanent Pasture to other
Areas in Cheshire in 1907

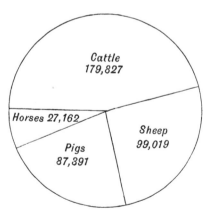

Fig. 8. Proportionate Numbers of Live Stock
in Cheshire in 1907